# PHARMACEUTICAL MATHEMATICS

Sara Byars

Cover image © Shutterstock.com

www.kendallhunt.com
*Send all inquiries to:*
4050 Westmark Drive
Dubuque, IA 52004-1840

Copyright © 2022 by Kendall Hunt Publishing Company

ISBN: 978-1-7924-7981-6

All rights reserved. No part of this publication may be reproduced,
stored in a retrieval system, or transmitted, in any form or by any means,
electronic, mechanical, photocopying, recording, or otherwise,
without the prior written permission of the copyright owner.

Published in the United States of America

Personal philosophy:

Too many students feel that they are "bad at math", and too many educators understand the solutions without being able to explain the process of arriving at the correct answer for math problems. This workbook was developed with those ideas in mind – to help students learn, and to help educators show the steps involved in the process of finding consistent solutions to all medical math questions. In this text, I focus on building foundational skills, and then surrounding those skills with sound logical processes and minimal memorization so that students can build solutions to any medical math question using these methods as their tools. With a significant focus on math literacy, students can use these tools with confidence, ease, and the knowledge that they will never again say that they are "bad at math".

# Contents

| | | |
|---|---|---|
| **Unit 1** | **Basic Mathematics** | **1** |
| Lesson 1 | Numbering Systems | 3 |
| Lesson 2 | Fractions | 13 |
| Lesson 3 | Decimals | 25 |
| Lesson 4 | Percent's | 37 |
| Lesson 5 | Ratio's | 43 |
| Lesson 6 | Proportions | 49 |
| Lesson 7 | Temperature Systems | 57 |
| Lesson 8 | Time | 61 |
| **Unit 2** | **Introduction to Units and Problem-Solving Methodology** | **73** |
| Lesson 9 | Systems of Measurement and the Metric System | 75 |
| Lesson 10 | The Household and Apothecary Measuring System | 83 |
| Lesson 11 | Understanding Math Word Problems (Math Literacy) | 93 |
| Lesson 12 | The Ratio/Proportion Method | 97 |
| Lesson 13 | Multistep Ratio/Proportion Problems | 105 |
| Lesson 14 | The Dimensional Analysis Method | 111 |
| Lesson 15 | Insulin and Other Unit Calculations | 119 |
| Lesson 16 | Compare and Contrast Methodology | 127 |
| **Unit 3** | **Retail Pharmacy Math** | **145** |
| Lesson 17 | Prescription Reading | 147 |
| Lesson 18 | Day Supply Calculations | 159 |
| Lesson 19 | Quantity Calculations | 167 |
| Lesson 20 | Refills and Total Pills | 175 |
| Lesson 21 | Measurements in the Lab | 185 |

| Lesson 22 | Accuracy and Percent Error | 197 |
| Lesson 23 | Nonsterile Compounding | 205 |

## Unit 4  Pharmacy Business Math ... 221

| Lesson 24 | Inventory Management | 223 |
| Lesson 25 | Income, Overhead, and Profit | 231 |
| Lesson 26 | Mark Up, Discounts, Depreciation | 239 |
| Lesson 27 | Patient Insurance Calculations | 247 |
| Lesson 28 | Pharmacy Insurance Calculations | 261 |

## Unit 5  Solution Calculations ... 275

| Lesson 29 | Understanding Solutions | 277 |
| Lesson 30 | Concentration Expression | 287 |
| Lesson 31 | Concentration Expression: Identities and Dilution Factors | 295 |
| Lesson 32 | Alligations | 305 |
| Lesson 33 | Solving Solution Problems | 315 |
| Lesson 34 | Math Literacy: Solution Problems | 329 |

## Unit 6  Hospital Pharmacy Math ... 345

| Lesson 35 | Powder Volume and Reconstitution | 347 |
| Lesson 36 | IV Push and Continuous IV Infusions | 363 |
| Lesson 37 | Flow Rate Calculations | 375 |
| Lesson 38 | Drop Factor Calculations | 383 |
| Lesson 39 | Flow Rate Variations | 395 |

## Unit 7  Pediatric Pharmacy Math ... 411

| Lesson 40 | Reference Text Pediatric Dose Calculations | 413 |
| Lesson 41 | Formula and BSA Pediatric Dose Calculations | 421 |
| Lesson 42 | Pediatric IV Administration | 431 |
| Lesson 43 | Daily Maintenance Needs | 441 |
| Appendix A | Math Concepts | 451 |
| Appendix B | Visual dosage chart for 30 days | 455 |
| Appendix C | Visual dosage chart for 4 weeks | 457 |

# UNIT 1
# Basic Mathematics

# LESSON 1

# Numbering Systems

Every number that you will encounter in pharmacy practice will be from one of two numbering systems: Roman or Arabic. The Arabic numbering system is by far the most common both inside and outside of healthcare, but occasionally pharmacy technicians will run across Roman numerals when interpreting a prescription.

## Roman Numerals:

What is the equivalent Arabic number for the following Roman numerals?

$\overline{ss}$ = _____ *the only Roman numeral that has 2 letters!

I, i, or ī = _____

V or v = _____

X or x = _____

L = _____

C = _____

D = _____

M = _____

## Rules for Interpreting a Roman Numeral:

1. Only compare \_\_\_\_\_ values at a time, in order from _____ to _____.

2. When a Roman numeral is repeated, the values are ( **added / subtracted** ) together.

   > **Example:** XXX -> Arabic numbers
   > X has a value of 10, so XXX = 10 + 10 + 10 => 30

   a. Roman numerals are never repeated more than ( **three / four** ) times in a sequence.

3. When comparing 2 Roman numerals that are in order from large to small, the values are ( **added** / **subtracted** )

> **Example**: VI -> Arabic number
>
> V has a value of 5, and I has a value of 1
>
> 5 is greater than 1, so the values of 5 and 1 are added together = 5 + 1 => 6

4. When comparing 2 Roman numerals that are in order from small to large, the values are ( **added** / **subtracted** )

> **Example**: IV -> Arabic numbers
>
> I has a value of 1, and V has a value of 5
>
> 1 is smaller than 5 so the values of 1 is subtracted from 5 = (1 – 5) => 4

*Note*: Roman numerals cannot be negative. To remind yourself of this, put all subtraction in parenthesis and find the positive difference between the values (sometimes called the absolute value).

5. A value can only be subtracted **once** from another value.

> **Example:** IIX is illogical; it is trying to represent the number 8 by using the following equation logic:
>
> (2-10). This makes sense in the Arabic System, but not in the Roman.
>
> Instead, 8 is written the following way: VIII (to represent 5 + 1 + 1 + 1 => 8)

6. There can be multiple subtractions within the same Roman numeral.

> **Example:** CDIX -> Arabic numbers
>
> (100 – 500) + (1 – 10) =
>
> 400 +      9      = 409

7. Roman numeral can only represent whole numbers with the exception of ½ (represented as s̄s̄)

## Conversion: Roman Numeral to Arabic Number

Step 1) Indicate the Arabic value below each Roman numeral.

> **Example:** Convert MMDXXIX to Arabic numbers.

**LESSON 1:** Numbering Systems

| M | M | D | X | X | I | X |
|---|---|---|---|---|---|---|
| 1,000 | 1,000 | 500 | 10 | 10 | 1 | 10 |

Step 2) Compare each value starting at the left side of the equation. If the values are the same, place a "+" sign between them. If they are different, ask the question, "Is this value larger than the value to the right?" If the answer is:

Yes = Place a "+" sign between the two values.

No = Place a "−" sign between the two values.

Move one value to the right and repeat the process for all the numerals in the sequence until you reach the end.

| M | | M | | D | | X | | X | | I | | X |
|---|---|---|---|---|---|---|---|---|---|---|---|---|
| 1,000 | + | 1,000 | + | 500 | + | 10 | + | 10 | + | 1 | − | 10 |

Step 3). Put parenthesis around any numbers that should be subtracted and perform this calculation first. Then, add your values together to solve! The result is the Arabic number equivalent.

1,000 + 1,000 + 500 + 10 + 10 + (1 − 10) = $\boxed{2,529}$

**STOP AND PRACTICE:** Convert the following Roman numerals to their Arabic number counterparts.

1. CX =   C   X

   Answer: _____

2. LIV =   L   I   V

   Answer: _____

3. IX =   I   X

   Answer: _____

4. DXIX =   D   X   I   X

   Answer: _____

5. MML =   M   M   L

   Answer: _____

## The Arabic Numbering System

The best way to convert from Arabic numbers to Roman numerals is to use _____ _____.

Indicate the values for the following number: 3,134.597

| Arabic numeral: | 3 | 1 | 3 | 4 | . | 5 | 9 | 7 |
|---|---|---|---|---|---|---|---|---|
| Place value (#): | | | | | | | | |
| Place value (name): | | | | | | | | |

Everything to the left of the decimal is a _____ number, and everything to the right of a decimal is a _____ number. Each time you move one value to the left, you _____ that value by _____, and each time you move between one value to the right, you _____ that value by _____.

**STOP AND PRACTICE:** Answer the following questions to discuss the concept of place value more.

1. List the value of each numeral in the named places for the number 92,375.481
    a. Tenth's place: _____
    b. Hundred's place: _____
    c. Thousand's place: _____
    d. Thousandth's place: _____
    e. Hundredth's place: _____

2. Determine the place value of the underlined digit.
    a. 18,<u>2</u>90.1: _____
    b. 5.<u>8</u>69: _____
    c. <u>6</u>1.2: _____
    d. 0.00<u>3</u>: _____
    e. 1.5<u>6</u>2: _____

## Conversion: Arabic Number to Roman Numeral

Convert the following numbers by following these steps:

Step 1) Split up the Arabic number into a sum of its place values.

# LESSON 1: Numbering Systems    7

**Example:** Convert 352 into Roman numerals.

```
      300
       50
  +     2
  _____
```

Step 2) Determine the Roman numeral equivalent of each place value. Remember to follow the rules we've already discussed!

```
      300  =  CCC
       50  =  L
  +     2  =  II
  _____
```

Step 3) Put the Roman numerals together in order (left to right) from the top to the bottom.

CCCLII

Step 4. (*Optional*) Convert your Roman numeral back to its Arabic number using the rules at the beginning of the lesson to check yourself!

**STOP AND PRACTICE:** Convert the following Arabic number's into Roman numerals.

1. 1,995

    Answer: _____

2. 47

    Answer: _____

3. 32

    Answer: _____

4. 129

    Answer: _____

5. 99

    Answer: _____

## Putting it All Together:

1. Indicate whether the value on the left is greater than (>), equal to (=) or lesser than (<).
   a. 5 _____ XI
   b. XX _____ 20
   c. 11 _____ IX
   d. 104 _____ CVI
   e. 1,900 _____ MCM

2. What is CXI – IV in roman numerals?

   Answer: _____

3. What is XXII + DV in roman numerals?

   Answer: _____

4. A prescription comes into the pharmacy that reads: "take ss tablet by mouth every day, dispense: XV"
   a. How many tablets are they taking per day?   Answer: _____
   b. How many tablets are being dispensed?   Answer: _____

5. A prescription comes into the pharmacy that reads: "take i-ii tablets every 4-6 hours, do not exceed VIII tablets per day, dispense: XXX"
   a. How many tablets can they take at one time?   Answer: _____
   b. How many tablets can they not exceed in one day?   Answer: _____
   c. How many tablets should be dispensed?   Answer: _____

## Sampling the Certification Exam:

1. What is XCVII in the Arabic numbering system?
   a. 97
   b. 907
   c. 102
   d. 107

   Answer: _____

2. Roman numerals written in descending order (biggest to smallest) are always:
   a. divided by each other
   b. added together
   c. multiplied by each other
   d. subtracted from each other

   Answer: _____

3. What is the place value of the digit 3 in the number: 434.291
   a. tens
   b. hundreds
   c. tenths
   d. hundredths

   Answer: _____

4. What number is in the tenths place in the number: 5.268
   a. 5
   b. 2
   c. 6
   d. 8

   Answer: _____

5. How would the number "one hundred and twelve and three tenths" be written using Arabic numbers?
   a. 112.310
   b. 112.3
   c. 11.23
   d. 0.1123

   Answer: _____

## Lesson 1 Content Check

1. Why is XXXX not the correct way to represent the number 40? _____
   _____
   _____
   _____

# UNIT 1: Basic Mathematics

   a. How should it be written using Roman numerals?

   Answer: _____

2. Why is LCV not the correct way to represent the number 55? _____

   a. How should it be written using Roman numerals?

   Answer: _____

3. Convert CDLXII into Arabic numbers.

   Answer: _____

4. Convert MCMIV into Arabic numbers.

   Answer: _____

5. Convert CD into Arabic numbers.

   Answer: _____

6. Convert CXII into Arabic numbers.

   Answer: _____

7. Convert DCXXIV into Arabic numbers.

   Answer: _____

8. Convert XLVII into Arabic numbers.

   Answer: _____

9. Perform the following: ivss – iii =

   Answer: _____

## LESSON 1: Numbering Systems

10. Perform the following: XLV + IX =

    Answer: _____

11. Perform the following: MCD divided by L =

    Answer: _____

12. Which of the following numbers indicates the thousandths place in the following number: 1,095.4928

    Answer: _____

13. Which number is larger for the following number: 56.351

    a. The digit in the tenths place or the digit in the tens place?

    Answer: _____

    b. The digit in the hundredths place or the digit in the thousandths place?

    Answer: _____

14. How would you write the number 99 using Roman numerals?

    Answer: _____

15. Convert 314 into roman numerals.

    Answer: _____

16. Convert 2,584 into roman numerals.

    Answer: _____

17. Convert 74 into roman numerals.

    Answer: _____

18. Convert 29 into roman numerals.

    Answer: _____

UNIT 1: Basic Mathematics

19. Convert 1,492 into roman numerals.

Answer: _____

20. Convert 522 into roman numerals.

Answer: _____

# LESSON 2

# Fractions

A fraction represents parts of a whole. The top number is called a **numerator** and represents the number of parts being defined. The bottom **number** is called a **denominator** and represents the total number of parts that the whole is split into. The line between two numbers of a fraction [ " – " or " / " ] symbolizes the mathematical function of division.

> **Example:** $\frac{1}{4}$ represents a whole split into 4 parts, and 1 of those parts is being defined by the fraction itself.
>
> Visually, this looks like this:
>
>

The larger the denominator, the smaller the parts. The larger the numerator, the more parts that are being represented. When fractions are being compared to one another in order to determine which is smaller or larger, the denominator must be the same to make an accurate comparison.

> **Example:** You have the following nitroglycerin tablets in stock in your pharmacy:
> 3/10 mg tablet, 4/10 mg tablet, and 6/10 mg tablet

The 3/10 mg tablet would be the smallest dose of nitroglycerin because only 3 parts of the whole that is cut into 10 are being represented by the fraction, whereas the 6/10 mg tablet represents 6 parts of the whole that is cut into 10, making it the largest dose of nitroglycerin.

Visually, each tablet can be represented like this:

3/10 mg tablet      4/10 mg tablet      6/10 mg tablet

# UNIT 1: Basic Mathematics

There are four types of fractions. Define and give an example of each:

| Type | Definition | Example |
|---|---|---|
| Proper | Numer less than Denominator | 5/10 |
| Improper | Numer > than Denom | 5/2, 8/5 |
| Mixed | Whole # + Proper fraction | 2 6/10 |
| Complex | Two fraction Bounded | 2/3 / 4/6 |

Follow-up:

1. **T / F** : The number 12 can be written as a fraction.
   a. How? __12/1__
   b. Why? __Anything divided by itself is itself__

2. **T / F** : It is sometimes necessary to convert a mixed number into an improper fraction.
   a. Why? __Yes for usual purpose__

3. **T / F** : Mixed numbers always convert to improper fractions, and improper fractions always convert to mixed numbers. __Yes__

## Conversion: Mixed Number to Improper Fraction

Convert mixed numbers into improper fractions by following these steps:

Step 1) Multiply the **whole number** by the **denominator**.

> **Example:** Convert $4 \& \frac{1}{2}$ into an improper fraction.

   4 is the **whole number** and 2 is the **denominator**.

   4 x 2 = 8

Step 2) Add the **numerator** to this number.

   1 is the **numerator**

   1 + 8 = 9

Step 3) Put the **sum** of step 2 over the existing **denominator**.

$$\frac{9}{2}$$

**LESSON 2:** Fractions    **15**

**STOP AND PRACTICE:** Convert the following mixed numbers into an improper fraction. Do not reduce or simplify them.

1. $7\frac{7}{15}$

    Answer: _112/15_

2. $6\frac{1}{3}$

    Answer: _19/3_

3. $1\frac{7}{8}$

    Answer: _15/8_

4. $5\frac{6}{13}$

    Answer: _71/13_

## Conversion: Improper Fraction to Mixed Number

Convert improper fractions into mixed numbers by following these steps:

Step 1) Divide the **numerator** by the **denominator**. Keep the number to the left of the decimal only – this becomes the whole number for your mixed number.

> **Example:** Convert $\frac{9}{2}$ into a mixed number

   9 is the **numerator** and 2 is the **denominator**

   9 divided by 2 = 4.5; therefore, your whole number is 4

Step 2) Multiply your whole number by the **denominator**, then subtract from your **numerator**. The difference becomes your **new numerator** for your mixed number.

   $4 \times 2 = 8$

   $9 - 8 = 1$

Step 3) The **denominator** for the mixed number stays the same. Put the whole number and your new fraction together to create a complete mixed number.

> 4 & 1/2

If you remember the process of long division, it is **exactly** the same.

**Example:**
$$2\overline{\smash{)}9} \quad \begin{array}{r} 4 \\ -8 \\ \hline 1 \text{ (remainder)} \end{array}$$

Two divides into nine four times evenly with a remainder of one left over, giving a fraction of 4 & 1/2

**STOP AND PRACTICE:** Convert the following improper fractions into mixed numbers. Do not reduce or simplify them.

1. $\dfrac{25}{2}$

    Answer: *12 1/2*

2. $\dfrac{14}{3}$

    Answer: *4 2/3*

3. $\dfrac{17}{5}$

    Answer: *3 2/5*

4. $\dfrac{91}{18}$

    Answer: *5 1/18*

## Functions of Fractions:

Performing addition and subtraction of fractions is a process that requires finding the lowest common denominator for each fraction because all denominators must be the same in order to perform either mathematical function. However, pharmacy technicians rarely add or subtract fractions more complicated than ½ or ¼, so this text will not cover these functions. In contrast, the concept of multiplication and division of fractions is relied on as a general foundation for future methodology in solving all pharmacy math problems. A general overview of how fractions will be presented in pharmacy math problems and applications will be discussed throughout the text.

**LESSON 2:** Fractions

**Multiplication of Fractions:**

Step 1) Multiply the **numerators** together – this becomes your new **numerator**.

Step 2) Multiply the **denominators** together – this becomes your new **denominator**.

**Example:** $\frac{1}{3} \times \frac{2}{5} = \frac{1 \times 2}{3 \times 5} = \frac{2}{15}$

**STOP AND PRACTICE:** Multiply the following fractions. Do not reduce or simplify them.

1. $\frac{1}{3} \times \frac{1}{5} =$   $1 \times 1 = 1$
   $3 \times 5 = 15$
   Answer: 1/15

2. $\frac{2}{7} \times \frac{4}{5} =$   $\frac{2 \times 4}{7 \times 5} = \frac{8}{35}$
   Answer: 8/35

3. $\frac{2}{5} \times \frac{6}{8} =$   $\frac{2 \times 6}{5 \times 8} = \frac{12}{14}$
   Answer: 12/40

4. $\frac{19}{2} \times \frac{4}{11} =$   $\frac{19 \cdot 4}{2 \cdot 11} = \frac{76}{22}$
   Answer: 76/22

## Dividing Fractions:

To divide fractions, use the acronym KCF to remember what to do.

KCF = Keep Change Flip

Step 1) Keep the first fraction

Step 2) Change the division sign to a multiplication sign

Step 3) Flip the second fraction; this is also known as finding the reciprocal, or mirror image, of the fraction.

Then, multiply the fractions as discussed above.

**Example:** $\frac{1}{4} \div \frac{2}{6} = \frac{1}{4} \times \frac{6}{2} = \frac{1 \times 6}{4 \times 2} = \frac{6}{8}$

Keep Change Flip

Follow-up:
1. T / F : Complex fractions are just a shorthand way of writing a division problem.
    a. Give an example.

**STOP AND PRACTICE:** Divide the following fractions. Do not reduce or simplify them.

1. $\frac{4}{8} \div \frac{6}{9} =$

    Answer: _____

2. $\frac{\frac{2}{7}}{\frac{3}{8}} =$

    Answer: _____

3. $\frac{11}{2} \div \frac{4}{12} =$

    Answer: _____

4. $\frac{\frac{3}{13}}{\frac{1}{6}} =$

    Answer: _____

## Reducing Fractions:

Reducing, or simplifying, a fraction is a process of representing a fraction in its lowest terms. When reducing, both the numerator and the denominator of the fraction must be divided evenly by the **same** whole number.

*Tips and Tricks:*

1. Always start with the smallest number you can divide both the numerator and the denominator **evenly** by. You can always repeat the process of division until no more matching dividers can be evenly done between the numerator and the denominator.
2. All **even** numbers can be divided by 2.
3. All numbers ending in 0 can be divided by 10.
4. All numbers ending in 5 or 0 can be divided by 5.

**LESSON 2:** Fractions

> **Example:** Reduce the fraction $\frac{4}{18}$ to its lowest terms.

Both the **numerator** and the **denominator** are divisible by 2.

$$\frac{4}{18} \div \frac{2}{2} = \boxed{\frac{2}{9}}$$

Follow-up:

1. **T / F :** Improper fractions can be reduced.
    a. Give an example.
2. **T / F :** Mixed numbers can be reduced.
    a. What is the catch to this? _____
    _____
    b. Give an example.
3. **T / F :** It is best to wait until the end of the problem to reduce fractions.
    a. Why? _____
    _____
    _____

**STOP AND PRACTICE:** Reduce the following fractions to their lowest value.

1. $\frac{3}{15}$

    Answer: _____

2. $\frac{4}{12}$

    Answer: _____

3. $\frac{5}{20}$

    Answer: _____

4. $\frac{6}{18}$

    Answer: _____

5. $\frac{72}{14}$

    Answer: _____

## Sampling the Certification Exam:

1. Which of these is a mixed fraction?

   a. $\frac{4}{5}$

   b. $\frac{12}{9}$

   c. $\frac{\frac{1}{2}}{\frac{3}{4}}$

   d. $1\frac{1}{2}$

   Answer: _____

2. What is the denominator in the fraction 2 & $\frac{10}{99}$?

   a. 10

   b. 99

   c. 2

   d. None of these

   Answer: _____

3. Calculate $\frac{14}{12} \times \frac{4}{5}$ and reduce to lowest terms.

   a. $\frac{56}{60}$

   b. $\frac{14}{15}$

   c. $\frac{70}{48}$

   d. $\frac{35}{24}$

   Answer: _____

4. Calculate $8 \div \frac{5}{6}$ and reduce to lowest terms.

   a. $\frac{48}{5}$

   b. $\frac{5}{48}$

   c. $\frac{40}{6}$

   d. $\frac{20}{3}$

   Answer: _____

5. Convert $5\frac{2}{7}$ to an improper fraction.

   a. $\frac{17}{7}$

   (b.) $\frac{37}{7}$

   c. $\frac{14}{7}$

   d. $\frac{33}{7}$

   Answer: _____

## Lesson 2 Content Check

*Please reduce all fractions from this point forward to get the correct answer!*

1. Convert $\frac{17}{9}$ to a mixed number.

   Answer: $1\frac{8}{9}$

2. Convert $102\frac{1}{4}$ to an improper fraction.

   Answer: 404/4

3. Convert $\frac{52}{10}$ to a mixed number.

   Answer: $5\frac{2}{10}$ or $\frac{1}{2}$

4. Convert $5\frac{3}{4}$ to an improper fraction.

   Answer: 23/4

5. Calculate $\frac{5}{3} \times \frac{8}{10}$ and reduce to lowest terms.

   Answer: 4/3

6. Calculate $\frac{6}{9} \times \frac{2}{6}$ and reduce to lowest terms.

   Answer: 1/2

UNIT 1: Basic Mathematics

7. Calculate $\frac{9}{7} \times \frac{7}{2}$ and reduce to lowest terms.

Answer: 9/2

8. Calculate $\frac{10}{16} \times \frac{5}{9}$ and reduce to lowest terms.

Answer: 25/72

9. Calculate $\frac{3}{8} \times \frac{4}{9}$ and reduce to lowest terms.

Answer: 1/6

10. Calculate $\frac{12}{9} \times 2$ and reduce to lowest terms.

Answer: 8/3

11. Calculate $\frac{5}{11} \div \frac{3}{4}$ and reduce to lowest terms.

Answer: 20/33

12. Calculate $\frac{4}{9} \div \frac{1}{2}$ and reduce to lowest terms.

Answer: 8/9

13. Calculate $\frac{1}{5} \div \frac{7}{11}$ and reduce to lowest terms.

Answer: 77/35

14. Calculate $\frac{2}{7} \div \frac{2}{6}$ and reduce to lowest terms.

Answer: 6/7

15. Calculate $\frac{6}{9} \div \frac{13}{4}$ and reduce to lowest terms.

Answer: 8/39

16. Calculate $\frac{3}{14} \div \frac{23}{36}$ and reduce to lowest terms.

Answer: 54/161

**LESSON 2:** Fractions

17. Reduce $\frac{68}{187}$ to the lowest terms.

    Answer: ~~68/187~~

18. Reduce $\frac{18}{3}$ to the lowest terms.

    Answer: 6

19. Reduce $\frac{24}{18}$ to the lowest terms.

    Answer: 4/3

20. Reduce $\frac{19}{2}$ to the lowest terms.

    Answer: 19/2

# LESSON 3

# Decimals

Place value is the most important consideration to make when discussing decimals. Fill in the following chart as a reminder of the place value for each digit:

| Digit | 3 | 1 | 3 | 4 | . | 5 | 9 | 7 |
|---|---|---|---|---|---|---|---|---|
| Place value (#): | *thousand* | *hundreds* | *tens* | *ones* | | *tenths* | *hundredths* | *thousandths* |
| Place value (name): | 1000 | 100s | 10s | 1s | | 1/10 | 1/100 | 1/1000 |

Where ___*whole*___ numbers are indicated to the left of the decimal and ___*personal*___ numbers are indicated to the right of the decimal.

Decimals should be read aloud by indicating the numerals behind the decimal as they are usually spoken, but adding the place value of the last digit to the right at the end to indicate its place within the numeral as a whole.

> **Example**: 13.21

Many people would say "thirteen point twenty-one", and most of the time, their audience would understand their meaning. However, the best way to speak it would be to say **"thirteen and twenty-one hundredths"**, eliminating any possible ambiguity that might exist.

**STOP AND PRACTICE:** Write out the correct way to say the numbers on the left in the space that follows.

1. 15.4 ___*fifteen and four tenths*___
2. 172.35 ___*one hundred and seventy two and thirty five*___
3. 0.46 ___*fourty six hundredths*___
4. 2.373 ___*two and three hundred and seventy three thousand*___

In healthcare, mistakes are easily made when writing down numbers that contain decimals. The most common errors are made when healthcare professionals forget to use **leading zeros,** or when they use **trailing zeros.**

**Leading zeros** are zeros that hold the place value before the decimal of a partial number.

> **Example:** "0.46" vs the incorrect ".46" or "0.1" vs the incorrect ".1"

In healthcare, they are _necessary_ because they communicate the exact value of the number being viewed without the potential for a misunderstanding. Without the zero in front of the decimal, the decimal itself can easily be seen as an accidental pen slip. For example, the numbers above could be misread as "46" or "1", which would cause a 10-fold increase in the value of the intended number.

**Trailing zeros** are zeros that follow the decimal of a whole number.

> **Example:** 1.0 or 20.0

In healthcare, they are unnecessary because they add no value to the number and can cause confusion. The addition of the zero behind the decimal casts doubt to the intention of the decimal itself – perhaps it was an accidental pen slip. For example, the numbers above could be misread as "10" or "200", which again, causes a 10-fold increase in the value of the intended number.

## Determination of Order:

Deciding the order from low to high, or high to low, of numerals with decimals is essential to the practice of pharmacy. Place value must be considered when determining the order of a series of numerals. An important piece of information to remember is that value increases the further to the left a digit is placed within a numeral.

> **Example:** Place the following numbers in order from largest to smallest: 0.03, 0.02, 0.033, 0.021, 0.31, ~~0.099~~ 0.99

Step 1) Place each numeral in vertical order so that each digit aligns by place value.

| 0 | . | 0 | 3 |   |
|---|---|---|---|---|
| 0 | . | 0 | 2 |   |
| 0 | . | 0 | 3 | 3 |
| 0 | . | 0 | 2 | 1 |
| 0 | . | 3 | 1 |   |
| 0 | . | 9 | 9 |   |

highest

Step 2) Fill in the blank spaces with zeros as this will not change the value of the number itself, but can be helpful in making determinations regarding order in the future.

**LESSON 3:** Decimals

| 0 | . | 0 | 3 | 0 |
|---|---|---|---|---|
| 0 | . | 0 | 2 | 0 |
| 0 | . | 0 | 3 | 3 |
| 0 | . | 0 | 2 | 1 |
| 0 | . | 3 | 1 | 0 |
| 0 | . | 9 | 9 | 0 |

Step 3) Starting at the place value furthest to the left, assign the numeral with the highest digit as the largest value. If two numerals have the same digits, move to the next place value to the right and compare – the highest digit will have the largest value. Continue until all numbers have been assessed.

| 0 | . | 0 | 3 | 0 |
|---|---|---|---|---|
| 0 | . | 0 | 2 | 0 |
| 0 | . | 0 | 3 | 3 |
| 0 | . | 0 | 2 | 1 |
| 0 | . | 3 | 1 | 0 |
| 0 | . | 9 | 9 | 0 |

The one's place does not have any digits, so looking at the place value to the right (the tenths place), we see that the numeral "0.99" has the largest digit (9), thus making it the largest number, followed by "0.31". The rest of the numerals have zeros in the tenths place, so we must move to the place value to the right (the hundredths place) to continue our determination of order.

| 0 | . | 0 | 3 | 0 |   |
|---|---|---|---|---|---|
| 0 | . | 0 | 2 | 0 |   |
| 0 | . | 0 | 3 | 3 |   |
| 0 | . | 0 | 2 | 1 |   |
| 0 | . | 3 | 1 | 0 | 2nd |
| 0 | . | 9 | 9 | 0 | 1st |

The bottom two numerals have already been placed in their consideration, so they can be ignored for the rest of the determination of order. In the hundredths place, two numerals have the digit 3, which is the highest value. Because they are the same, we must move to the place value to the right (the thousandths place) to continue our determination of order.

## Unit 1: Basic Mathematics

| 0 | . | 0 | 3 | 0 |
|---|---|---|---|---|
| 0 | . | 0 | 2 | 0 |
| 0 | . | 0 | 3 | 3 |
| 0 | . | 0 | 2 | 1 |
| 0 | . | 3 | 1 | 0 | 2nd
| 0 | . | 9 | 9 | 0 | 1st

The numeral "0.033" has the largest digit (3), thus making it the next number in order, followed by the numeral "0.03".

The process can then be repeated with the rest of the numbers remaining:

| 0 | . | 0 | 3 | 0 | 4th
|---|---|---|---|---|
| 0 | . | 0 | 2 | 0 |
| 0 | . | 0 | 3 | 3 | 3rd
| 0 | . | 0 | 2 | 1 |
| 0 | . | 3 | 1 | 0 | 2nd
| 0 | . | 9 | 9 | 0 | 1st

Once again, two numerals have the digit 2, which is the only remaining value. Because they are the same, we must move to the place value to the right (the thousandths place) to continue determination of order. The numeral "0.021" has the largest digit (1), thus making it the next number in the order, followed by the numeral "0.02".

> Final result: the numerals in order from largest to smallest are 0.99, 0.31, 0.033, 0.03, 0.021, 0.02

Therefore: The place value closest to the decimal with the highest numerical value is the **(largest / smallest)**

**STOP AND PRACTICE**: Place the following series in order from largest to smallest.

1. 1.21, 1.33, 1.45, 1.05, 1.23

Answer: _____

2. 0.101, 0.144, 0.1, 0.213, 0.011

Answer: _____

3. 3.22, 3.45, 3.04, 3.99, 3.09, 3.091

Answer: _____

4. 14.001, 14.005, 14.05, 14.061, 14.52

Answer: _____

# Rounding:

Rounding a number to a specific place value is an attempt to abbreviate that number by determining if it is closer to the lower or the higher number of that same place value.

> **Example**: Round 6.1 to the nearest one's place.

This question is *essentially* asking if 6.1 is closer to the number 6 or the number 7 – all numbers whose value is in the one's place where the low end is kept the same, and the higher end is the next value in line. If you picture it on a number line, it becomes easy to see:

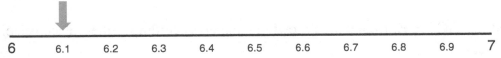

As you can see, the number 6.1 is closer to 6 than it is to 7, so 6.1 rounded to the nearest one's place would be 6.

> **Example:** Round 17.54 to the nearest ten's place.

This question is *essentially* asking if 17.54 is closer to the number 10 or the number 20 – all numbers whose value is in the ten's place where the low end is kept the same, and the higher end is the next value in line. On a number line, it would look like this:

# Unit 1: Basic Mathematics

As you can see, the number 17.54 is clearly closer to 20 than it is to 10, so 17.54 rounded to the nearest ten's place would be 20.

**Rules:**

Step 1) Identify the place value that you are rounding.

> **Example:** Round the number 4.589 to the **tenth's** place.
>
> The number 5 is in the tenths place.

Step 2) Look at the digit to the RIGHT to determine what to do next.

4.589

Step 3) If the value of the digit in consideration is 4 or less, _do nothing_.
If the value of the digit in consideration is greater than 5, _move original up by one_.

    8 > 5 so, the number in consideration (5) rounds UP to the next digit in line (to 6).

Step 4) Replace the rest of the digits to the right with zeros, or drop the same digits in the case of partial numbers where additional zeros would not be appropriate.

> **4.6** is your rounded number.

**STOP AND PRACTICE:** Round the following numbers according to their place value.

1. 143.684
   a. To the nearest tenth's place: _143.7_
   b. To the nearest hundredth's place: _143.68_
   c. To the nearest hundred's place: _100_
   d. To the nearest one's place: _144_
   e. To the nearest ten's place: _140_

2. 1,599.898
   a. To the nearest hundredths place: _1599.9_
   b. To the nearest tenths place: _1599.9_
   c. To the nearest ones place: _1600_
   d. To the nearest tens place: _1600_
   e. To the nearest hundreds place: _1600_
   f. To the nearest thousands place: _2000_

What do you notice about #2 in the STOP AND PRACTICE section? _____
_____
_____

1. Therefore, what is the rule regarding rounding when the **place value to be rounded** is the number 9? _____
_____
_____

2. How does rounding seem to affect accuracy to the original value? _____
_____
_____
_____

3. T / F : Always wait to round until the end of a problem.

    a. Why? _____
    _____
    _____

## Conversions:

1. T / F : Fractions and decimals represent the same thing.

    To convert from a fraction to a decimal:

    Step 1) Make sure your fraction is a proper or improper fraction. If it is not, convert it.

    Step 2) Divide the numbers in your calculator and record the answer.

    > **Example:** ½ is equivalent to 1 divided by 2 = 0.5

    To convert from a decimal to a fraction:

    Step 1) Determine the place value of the digit that is furthest to the right of the decimal. The Arabic number equivalent of that place value becomes your denominator.

    > **Example:** 0.113 -> fraction =
    >
    > 3 is the furthest digit to the right. It is in the thousandths place value. Therefore, our denominator is 1,000.

    Step 2) The digits behind the decimal become the numerator. Put these digits over your denominator.

    $$\frac{113}{1000}$$

Step 3) Reduce if necessary.

**\*NOTE:\*** If there is a whole number to the left of your decimal, it then becomes a mixed fraction.

> **Example:** 2.113 -> fraction = 2 & $\frac{113}{1000}$

**STOP AND PRACTICE:** Convert the following. Reduce when necessary and round to the nearest thousandth place:

1. Convert 0.23 into a fraction.

    Answer: __23/100__

2. Convert 0.03 into a fraction.

    Answer: __3/100__

3. Convert 10.17 into a fraction.

    Answer: __10 $\frac{17}{100}$__

4. Convert $\frac{5}{1004}$ into a decimal.

    Answer: __0.005__

5. Convert 11 & $\frac{7}{8}$ into a decimal.

    Answer: _____

## Sampling the Certification Exam:

1. Which digit is in the tens place in the number 4,892.103?
    a. 4
    b. 8
    c. 1
    d. 9

    Answer: _____

2. Which of the following best represents the number "five and twenty-seven thousandths"?

   a. 5.27

   b. 5.027

   c. 5.0027

   d. 5.00027

   Answer: _____

3. What is the smallest numeral in the series: 0.09, 0.091, 0.009, 0.99

   a. 0.09

   b. 0.091

   c. 0.009

   d. 0.99

   Answer: _____

4. Round the following number to the nearest hundredths place: 1.2268

   a. 1.0

   b. 1.2

   c. 1.23

   d. 1.227

   Answer: _____

5. What is 9.81 ÷ 1.03 rounded to the nearest hundredths place?

   a. 9.5

   b. 9.5243

   c. 9.524

   d. 9.52

   Answer: _____

## Lesson 3 Content Check

1. Which number is larger? 0.34 or 0.056

   Answer: _____

2. Write out the following as both a decimal and a fraction: nineteen and five sevenths.

Answer: _____

3. What is 6.5543 + 19.5799 rounded to the nearest tenths place?

Answer: _____

4. Convert $\frac{4}{63}$ into a decimal.

Answer: _____

5. Convert $\frac{1}{22}$ into a decimal and round to the nearest thousandths place.

Answer: _____

6. Convert $\frac{2}{394}$ into a decimal.

Answer: _____

7. Convert 1.3 to a fraction.

Answer: _____

8. Put these values in order of smallest to largest: 0.04, 0.084, 0.03, 0.10, 0.106

Answer: _____

9. Write out the following as both a decimal and a fraction: ninety-seven thousandths.

Answer: _____

10. Convert 0.694 into a fraction.

Answer: _____

11. Convert 0.34 to a fraction.

Answer: _____

**LESSON 3:** Decimals

12. Convert 0.002 into a fraction.

    Answer: _____

13. **T / F :** 0.7 is the same as 0.70

    Answer: _____

14. **T / F :** Adding a zero to the end of a number after a decimal changes its value.

    Answer: _____

15. Round the number 5,920.019 to the nearest thousands place.

    Answer: _____

16. Round the number 91,999.99 to the nearest hundredths place.

    Answer: _____

17. Round the number 111.111 to the nearest one's place.

    Answer: _____

18. Which of the following best represents a number with a leading zero: 0.46, 10.1, 100.0

    Answer: _____

19. Write out the following as both a decimal and a fraction: four and seven tenths

    Answer: _____

20. Write out the following as both a decimal and a fraction: seventy-seven and seven hundred and two thousandths

    Answer: _____

## LESSON 4

# Percent's

The term percent means "out of one hundred" and can be represented by the symbol "%". In pharmacy practice, percent's are often seen in concentrated or diluted solutions, topical formulations and in the hospital setting through IV's. The symbol and meaning of the word helps to give value to a number that otherwise has no meaning beyond the numeral itself.

## Conversions:

To convert from a percent to a fraction:

Step 1) Drop the % sign, and put the number over 100. Reduce if necessary. (Seriously, that's it!)

> **Example:** 30% -> fraction
>
> $\frac{30}{100}$ which reduces to $\frac{3}{10}$ by dividing both the numerator and the denominator by 10.

To convert from a percent to a decimal:

Step 1) Drop the % sign and divide by 100. <u>Don't forget a leading zero!</u>

> **Example:** 30% -> decimal
>
> $30 \div 100 = 0.3$

To convert from either a decimal or a fraction to a percent requires the use of proportions, which is discussed later in the text (lesson #6).

**STOP AND PRACTICE:** Convert the following. Reduce when necessary and round to the nearest thousandth place:

| Percent | Fraction | Decimal |
|---|---|---|
| 27% | 27/100 | 0.27 |
| 80% | 80/100 | 0.8 |
| 4% | 4/100  1/25 | 0.04 |
| 52% | 52/100 = 13/25 | 0.52 |

1. **T / F :** It is possible to have a decimal within a fraction.
    a. What can you do if you don't want a decimal within a fraction? _____
    _____
    _____
        i. Prove it! Convert 0.3% as a fraction with NO decimals.

                                                                Answer: _____

2. **T / F :** 0.7% is the same as 0.7.
    a. Why/why not? _____
    _____
    _____

    b. Convert 0.9% to a decimal.

                                                                Answer: _____

## Sampling the Certification Exam:

1. Convert 4% to a fraction and reduce to the lowest terms.
    a. $\frac{4}{100}$
    b. $\frac{2}{50}$
    c. $\frac{1}{25}$
    d. $\frac{4}{10}$

                                                                Answer: _____

2. Convert 1.5% to a decimal.
    a. 0.015
    b. 0.15
    c. 1.5
    d. 0.0015

                                                                Answer: _____

LESSON 4: Percent's

3. The term percent means:
   a. Over 100
   b. Out of 100
   c. Pertaining to 100
   d. Compared to 100

   Answer: _____

4. Which of the following does NOT represent 2.54% as a fraction?
   a. $\dfrac{2.54}{100}$
   b. $\dfrac{25.4}{1,000}$
   c. $\dfrac{254}{10,000}$
   d. $\dfrac{25.4}{100}$

   Answer: _____

5. A bag of normal saline is always 0.9%. What is this percent represented as a decimal?
   a. 0.9
   b. 0.09
   c. 0.009
   d. 0.0009

   Answer: _____

## Lesson 4 Content Check

1. Convert 65% to a fraction and reduce to lowest terms.

   Answer: _____

2. Convert 1.7% to a fraction and reduce to lowest terms.

   Answer: _____

3. Convert 20% to a fraction and reduce to lowest terms.

   Answer: _____

4. Convert 16.39% to a fraction and reduce to lowest terms.

   Answer: _____

5. Convert 4.922% to a fraction and reduce to lowest terms.

   Answer: _____

6. Convert 94% to a fraction and reduce to lowest terms.

   Answer: _____

7. Convert 67.53% to a fraction and reduce to lowest terms.

   Answer: _____

8. Convert 10.001% to a fraction and reduce to lowest terms.

   Answer: _____

9. Convert 13.02% to a fraction and reduce to lowest terms.

   Answer: _____

10. Convert 0.08% to a fraction and reduce to lowest terms.

    Answer: _____

11. Convert 17% to a decimal.

    Answer: _____

12. Convert 0.02% to a decimal.

    Answer: _____

13. Convert 76.2% to a decimal.

    Answer: _____

14. Convert 0.09% to a decimal.

    Answer: _____

**LESSON 4:** Percent's

15. Convert 10.01% to a decimal.

   Answer: _____

16. Convert 8% to a decimal.

   Answer: _____

17. Convert 46.58% to a decimal.

   Answer: _____

18. Convert 99.1% to a decimal.

   Answer: _____

19. Convert 26.003% to a decimal.

   Answer: _____

20. Convert 14% to a decimal.

   Answer: _____

# LESSON 5

# Ratio's

Ratios are comparisons of one value to another and are represented mathematically as A : B, where the colon in a ratio can be read as **per, out of, to,** or **in**. The problems themselves usually define these further, and these definitions and their meaning will be discussed later in the text.

> **Examples:**
> A 1:2 alcohol and water solution can be thought of as 1 part of alcohol **per** 2 parts of water.
> A 1:4 insulin and water solution can be thought of as 1 part of insulin **out of** 4 parts of water.
> A 1:100 nitroglycerin and sodium chloride solution can be thought of as 1 part of nitroglycerin **to** 100 parts of sodium chloride.
> A 1:2,000 bleach and water solution can be thought of as 1 part bleach **in** 2,000 parts water.

1. What is the difference between a ratio and a fraction? _____
2. T / F : Ratio's represent parts unless given meaning by adding a unit.

   a. Explain: _____
   _____
   _____

## Conversions:

To convert from a fraction to a ratio:

Step 1) Change the "/" symbol to a ":" symbol. Reduce if necessary.

> **Example:** ½ -> ratio
> 1 : 2

43

To convert from a ratio to a fraction:

Step 1) Change the ":" symbol to a "/" symbol. Reduce if necessary.

> **Example:** 1 : 2 -> fraction
> ½

To convert from a ratio to a decimal:

Step 1) Convert to a fraction, and then follow the rules for converting a fraction to a decimal.

**STOP AND PRACTICE:** Convert the following and reduce to the lowest terms:

1. 2 : 12 -> fraction

    Answer: _____

2. 6 : 50 -> fraction

    Answer: _____

3. 0.6 -> ratio

    Answer: _____

4. ¾ -> ratio

    Answer: _____

5. $\frac{4}{13}$ -> ratio

    Answer: _____

## Sampling the Certification Exam:

1. A recipe for a compound calls for 1 parts of alcohol to 2 parts of water. What would this look like as a ratio?

    a. 1 : 3
    b. 1 : 4
    c. 2 : 3
    d. 1 : 2

    Answer: _____

2. All ratios have which symbol between the numbers being compared?

   a. ;

   b. –

   c. /

   d. :

   Answer: _____

3. Reduce $\frac{35}{190}$ to the lowest terms and express as a ratio.

   a. 5 : 27

   b. 7 : 38

   c. 27 : 5

   d. 38 : 7

   Answer: _____

4. Express 0.63 as a ratio.

   a. 0.63 : 10

   b. 10 : 0.63

   c. 63 : 100

   d. 100 : 63

   Answer: _____

5. A prescriber says to mix a one-to-three-part solution of alcohol to water. What would this look like as a ratio?

   a. 1 : 3

   b. 1 : 4

   c. 1 : 2

   d. 3 : 1

   Answer: _____

## Lesson 5 Content Check

1. Convert 4 : 62 to a fraction and reduce to lowest terms.

   Answer: _____

2. Reduce and convert $\frac{75}{315}$ to a ratio.

Answer: _____

3. Convert 18 : 201 to a fraction and reduce to lowest terms.

Answer: _____

4. Convert 15 : 300 to a fraction and reduce to lowest terms.

Answer: _____

5. Reduce and convert $\frac{24}{96}$ to a ratio.

Answer: _____

6. Reduce and convert $\frac{40}{200}$ to a ratio.

Answer: _____

7. Reduce and convert $\frac{112}{2004}$ to a ratio.

Answer: _____

8. Convert 4 : 50 into a decimal.

Answer: _____

9. Convert 3 : 5 into a fraction and reduce to lowest terms.

Answer: _____

10. Convert 1 : 20 into a decimal.

Answer: _____

11. Convert 0.25 into a ratio.

Answer: _____

12. Convert 3 : 287 into a decimal.

Answer: _____

**LESSON 5:** Ratio's

13. Reduce and convert $\frac{9}{81}$ to a ratio.

    Answer: _____

14. Write out the ratio for the following mixture: 1 part lemonade to 3 parts tea.

    Answer: _____

15. Write out the ratio for the following mixture: 4 parts bleach to 10 parts water.

    Answer: _____

16. Write out the ratio for the following mixture: 1 part vinegar in 20 parts water.

    Answer: _____

17. Convert 0.008 into a ratio.

    Answer: _____

18. Convert 14 : 987 into a decimal.

    Answer: _____

19. Convert $\frac{0.3}{2}$ to a ratio.

    Answer: _____

20. Convert $\frac{9.98}{1001}$ to a ratio.

    Answer: _____

# LESSON 6

# Proportions

Proportions are two ratios, or fractions, that are equal in value. They can be represented as two ratios separated by a double colon (::) or two fractions separated by an equal sign ( = ).

**Example:** $\frac{1}{3} = \frac{4}{12}$ or 1 : 3 :: 4 : 12 Either can be read as "one is to three as four is to twelve"

In order to determine if a proportion is truly equal, use cross multiplication diagonally across the equal sign for both numerators and compare the end result. If the cross products are the same number, the proportion is considered to be a "true" proportion.

**Example:** $\frac{1}{3} = \frac{4}{12}$

$3 \times 4 = \quad 1 \times 12 =$
$12 \quad = \quad 12$

**STOP AND PRACTICE:** Determine whether or not the following are true proportions:

1. 3 : 4 :: 12 : 48

    Answer: _____

2. $\frac{5}{10} = \frac{60}{120}$

    Answer: _____

3. $\frac{10}{40} = \frac{2}{20}$

    Answer: _____

4. 5 : 30 :: 20 : 100

    Answer: _____

## Solving Proportions:

1. **T / F :** In order to solve a proportion, 3 of the 4 values must be known.

2. How are proportion problems solved? _____

   a. Describe in your own terms: _____
   
   _____
   
   _____

   b. Give an example.

**STOP AND PRACTICE:** Using cross multiplication, solve for the following missing values:

1. $3 : 5 :: 5 : X$

   Answer: _____

2. $\dfrac{40}{60} = \dfrac{X}{75}$

   Answer: _____

3. $X : 15 :: 100 : 300$

   Answer: _____

4. $\dfrac{95}{X} = \dfrac{45}{2}$

   Answer: _____

Follow up: Did it matter where the $X$ was in any of the above equations? _____

## Conversions:

To convert from a percent to a ratio:

Step 1) Change the symbol "%" to a " : " and then add the number 100 afterwards (because percent means "out of 100"). Reduce if necessary.

> **Example:** Convert 30% to a ratio
>
> 30 : 100 where both are divisible by 10, so 3 : 10 would also be acceptable

NOTE: Most ratios in medical math have the number 1 to the left of a colon.

> **Example:** Your pharmacy stocks a 1:40 Lugol's solution on its shelf.

**LESSON 6:** Proportions

If you come up with an answer that does not have a 1 to the left of the colon, a simple proportion can be used to calculate the ratio by setting the answer equal to $\frac{1}{X}$, solving, and reporting the answer as $1 : X$

> **Example:** Determine the ratio of the fraction $\frac{4}{200}$.
>
> $\frac{4}{200} = \frac{1}{X}$ where $X = 50$ so the answer is 1:50

To convert from a decimal, fraction, OR ratio to a percent:

Step 1) Start with a fraction. If it is not, convert it.

Step 2) Set up a proportion equal to $\frac{X}{100}$

Step 3) Cross multiply and divide.

Step 4) Add a % sign to the answer you get for $X$.

> **Example:** Convert 0.4 to a percent
>
> $\frac{4}{10} = \frac{X}{100}$ where $x = 40$, so the answer is 40%

> **Example:** Convert ½ to a percent
>
> $\frac{1}{2} = \frac{X}{100}$ where $x = 50$, so the answer is 50%

> **Example:** Convert 3:20 to a percent
>
> $\frac{3}{20} = \frac{X}{100}$ where $x = 15$, so the answer is 15%

**STOP AND PRACTICE:** Convert the following. Reduce when necessary and round to the nearest thousandth place:

| Percent | Fraction | Decimal | Ratio |
|---|---|---|---|
| 0.2% | | | |
| | 4/31 | | |
| | | 0.697 | |
| | | | 1:4000 |

## Using Percent's in Math Problems:

Proportions are your friend! Remember, percent ALWAYS means out of 100, which can be represented mathematically as $X\% = \frac{X}{100}$.

1. Any number following the word "of" is the denominator of the fraction. As a problem, it might look like "X% of B" where **B** = total number; or the **denominator** of the unknown fraction

   > **Example:** 10% of 70
   >
   > $\frac{10}{100} = \frac{X}{70}$ where $X = 7$

2. Any number before or immediately following the word "is" is the numerator of the fraction. As a problem, it might look like "X% of what number is A" where **A** = the partial number; or the **numerator** of the unknown fraction

   > **Example:** 7% of what number is 91?
   >
   > $\frac{7}{100} = \frac{91}{X}$ where $X = 1300$

3. Other problems won't give a percent within the problem but will instead, ask what the percent is. As a problem, it might look like "**A** is what percent of **B**", and the same logic as above can be used to read, set up, and then solve the problem

   > **Example:** 5 is what percent of 50?
   >
   > $\frac{5}{50} = \frac{X}{100}$ where $X = 10\%$

**STOP AND PRACTICE:** Calculate the following:

1. 21 is what percent of 159?

   Answer: _____

2. 22% of what number is 77?

   Answer: _____

3. What is 15% of 19?

   Answer: _____

4. 13% of what number is 21?

   Answer: _____

# LESSON 6: Proportions

## Sampling the Certification Exam:

1. 21 is what percent of 159?
    a. 13%
    b. 33%
    c. 35%
    d. 757%

    Answer: _____

2. 35 is what percent of 212?
    a. 6%
    b. 15%
    c. 17%
    d. 74%

    Answer: _____

3. Solve for $X$ in the following problem - 250 : 1000 :: $X$ : 4
    a. 1
    b. 16
    c. 2
    d. 8

    Answer: _____

4. 22% of what number is 77?
    a. 17
    b. 29
    c. 350
    d. 375

    Answer: _____

5. What is 0.4% of 150?
    a. 0.06
    b. 6
    c. 0.6
    d. 60

    Answer: _____

## Lesson 6 Content Check

1. Solve for $X$ and round to the nearest hundredth place: $X : 12 :: 54 : 412$

    Answer: _____

2. Solve for $X$ and round to the nearest ones place: $\dfrac{X}{10} = \dfrac{20}{40}$

    Answer: _____

3. Solve for $X$ and round to the nearest tens place: $\dfrac{55}{X} = \dfrac{900}{2560}$

    Answer: _____

4. Solve for $X$ in the following proportion: $\dfrac{X}{4} = \dfrac{15}{60}$

    Answer: _____

5. Solve for $X$ in the following proportion: $3 : X :: 7 : 21$

    Answer: _____

6. Solve for $X$ in the following proportion: $\dfrac{6}{72} = \dfrac{X}{9}$

    Answer: _____

7. Solve for $X$ in the following proportion: $0.2 : 1.3 :: 4.6 : X$

    Answer: _____

8. Is the following a true proportion? $\dfrac{5}{6} = \dfrac{30}{36}$

    Answer: _____

9. Solve for $X$ in the following proportion: $0.8 : 10 :: X : 2$

    Answer: _____

10. Is the following a true proportion? $\dfrac{4}{9} = \dfrac{37}{84}$

    Answer: _____

LESSON 6: Proportions

11. 35 is what percent of 212?

    Answer: _____

12. What is 9% of 70?

    Answer: _____

13. 10% of what number is 40?

    Answer: _____

14. 15 is what percent of 68?

    Answer: _____

15. 30 is what percent of 1,597?

    Answer: _____

16. Calculate the following test score to the nearest tenth of a percent: 35 out of 75 questions correct.

    Answer: _____

17. What is 0.5% of 25?

    Answer: _____

18. 12% of what number is 100?

    Answer: _____

19. What is 15% of $189.99?

    Answer: _____

20. 8.25 is what percent of 60?

    Answer: _____

# LESSON 7

# Temperature Systems

The two most common temperature systems used in pharmacy are Fahrenheit and Celsius. Both are used based on the freezing and boiling point of water.

Fill out the following chart to have as a comparison:

| Level | Fahrenheit (°F) | Celsius (°C) |
|---|---|---|
| Freezing | | |
| Boiling | | |

## Converting Between Systems

Only one formula needed:

> $9C = 5F - 160$ where C stands for Celsius and F stands for Fahrenheit

Plug the value given into the equation and use basic algebra to solve.

*Tips and Tricks:*

1. In order to eliminate a variable or digit within an equation, you must perform its opposite function. For this to work properly think of the following as equal but opposite functions:

    a. Addition and subtraction

    b. Multiplication and division

2. What you do to one side of the equation, you must do to the other to keep it balanced. The key is in the equal sign!

3. Pay attention and don't forget about the order of operations. This is a method that gives instructions on the order in which to solve a problem. Fill in the following chart

to serve as a helpful mnemonic, or memory device, to remembering the order of operations:

| Mnemonic | Please | Excuse | My | Dear | Aunt | Sally |
|---|---|---|---|---|---|---|
| Acronym | P | E | M | D | A | S |
| Meaning | | | | | | |

*Note* M/D and A/S are on the same level of order as one another. If a problem has both multiplication and division, or addition and subtraction within the same level of order, perform the calculations from left to right.

## Using the Formula:

Let's prove the chart above.

1. If given Celsius

   **Example:** Convert 0°C to °F

   | | |
   |---|---|
   | $9(0) = 5F - 160$ | Plug in 0 in the C spot since it is given in Celsius |
   | $9 \times 0 = 5F - 160$ | A variable sitting next to a digit implies multiplication. Perform the multiplication first (according to PEMDAS) |
   | $0 = 5F - 160$ $+160 = +160$ $160 = 5F - 0$ | Eliminate the "−160" since there are no variables attached to it; to get rid of something that was subtracted, we have to add it back to BOTH sides of the equation |
   | $\dfrac{160}{5} = \dfrac{5F}{5}$ | "5F" implies multiplication. To get rid of the 5 we have to do the opposite (divide) |
   | $32° = F$ | Make sure to put the degree symbol! |

2. If given Fahrenheit

   **Example:** Convert 212°F to °C

   | | |
   |---|---|
   | $9C = 5(212) - 160$ | Plug in 212 in the F spot since it is given in Fahrenheit |
   | $9C = (5 \times 212) - 160$ | Multiply (according to PEMDAS) |
   | $9C = 1,060 - 160$ | Perform the subtraction (next step, according to PEMDAS) |
   | $\dfrac{9C}{9} = \dfrac{900}{9}$ | "9C" implies multiplication. To get rid of the 9 we have to do the opposite (divide) |
   | $C = 100°$ | Make sure to put the degree symbol! |

## Sampling the Certification Exam:

1. A newborn's temperature reads 102.4°F, which is _____ °C.
   a. 149.8°C
   b. 213.8°C
   c. 74.6°C
   d. 39.1°C

   Answer: _____

2. A vial of medroxyprogesterone should be stored at 20°C, or _____ °F.
   a. 37.5°F
   b. 28.9°F
   c. 88.7°F
   d. 68°F

   Answer: _____

3. What is -14°F converted to Celsius?
   a. 32.3°C
   b. -18.3°C
   c. 1.2°C
   d. -25.6°C

   Answer: _____

4. A storage room is kept at 8°C, which is the same as _____ °F.
   a. 19.2°F
   b. 28.9°F
   c. -13.3°F
   d. -4.1°F

   Answer: _____

5. The fridge in the pharmacy reads at 12.5°C, or _____ °F.
   a. 54.5°F
   b. 48.2°F
   c. 23.9°F
   d. 62°F

   Answer: _____

# UNIT 1: Basic Mathematics

## Lesson 7 Content Check

1. Convert 42°C to Fahrenheit.

   Answer: _____

2. Convert 55°F to Celsius.

   Answer: _____

3. Convert 14°F to Celsius.

   Answer: _____

4. Convert 60°C to Fahrenheit.

   Answer: _____

5. Convert 25°F to Celsius.

   Answer: _____

6. Convert 13°C to Fahrenheit.

   Answer: _____

7. Convert 98.7°F to Celsius.

   Answer: _____

8. Convert 13.8°C to Fahrenheit.

   Answer: _____

9. Convert 182°F to Celsius.

   Answer: _____

10. Convert 2°F to Celsius.

    Answer: _____

# LESSON 8

# Time

There are two systems used to tell time – the AM/PM (aka: the 12-hour method), or military time. Most people outside of the military are more familiar with the 12-hour method as most clocks use that as their display. The AM or PM designation listed next to the numbers of the time being displayed has a significance and a meaning. AM stands for ante meridiem, which is Latin for "before midday", and PM stands for post meridiem, which is Latin for "after midday". In both, midday is another term used interchangeably for noon, or 12:00 PM. Therefore, the abbreviations are designed to indicate the part of the day being referenced. All time communicated to patients should be in AM/PM time.

With military time, there is no ambiguity on when exactly something occurred or is being referenced to. The digits used to display the time, always in a four-digit sequence, are telling of exactly when something occurred as there is only one 13th hour of the day, or one 17th hour of the day, etc. In military time, midnight is considered the start of the day, just as in AM/PM time where midnight is the time in which the calendar would flip to the next day in sequence. This start is indicated as 0000 (sometimes written as 00:00), and from there, the time continues to count upwards towards the 24-hour mark. There are still only 60 minutes in an hour, so the largest value that can occur in the last two digits is 59.

Follow-up: What area of pharmacy might military time be used the most frequently? _____

## Conversion – AM/PM to Military Time:

1. Midnight – 12:59 AM = keep minutes; drop AM; change hours to 00
2. 1:00 AM – 12:59 PM = keep as is; drop AM/PM; make 4 digits by adding a zero in the front, if needed.
3. 1:00 PM – 11:59 PM = add 12 hours; drop AM/PM; make 4 digits by adding a zero in the front, if needed.

**STOP AND PRACTICE:** Convert the following to military time:

1. 7:30 am

    Answer: _____

2. 4:28 pm

    Answer: _____

3. 12:45 am

    Answer: _____

4. 9:20 pm

    Answer: _____

5. 2:24 am

    Answer: _____

## Conversion – Military Time to AM/PM:

1. 0000 – 0059 = change first two digits to 12; keep last two digits; add AM to the end.
2. 0100 – 1159 = keep as is; add AM to the end.
3. 1200 – 1259 = keep as is; add PM to the end.
4. 1300 – 2359 = subtract 1200; add PM to the end.

**STOP AND PRACTICE:** Convert the following military time into standard am/pm time:

1. 17:30

    Answer: _____

2. 23:49

    Answer: _____

3. 15:22

    Answer: _____

4. 00:34

    Answer: _____

5. 1204

    Answer: _____

Outside of these time conversions, patients often use colloquialisms to indicate a certain time that a good pharmacy technician should be familiar with, such as:

- "A quarter past the hour" or "A quarter until the hour" – this means 15 minutes after, or 15 minutes before, the hour of time indicated, respectively. In this case, patients are referring to the fact that there are 4 "quarters" of time within an hour. If one hour is 60 minutes, one-quarter $\left(\frac{1}{4}\right)$ of one hour equals to 15 minutes.

- "One-half past the hour" or "One-half hour before" – this means 30 minutes after, or 30 minutes before, the hour of time indicated, respectively. In this case, patients are referring to the fact that there are 2 "halves" of time within an hour. If one hour is 60 minutes, one-half $\left(\frac{1}{2}\right)$ of one hour equals to 30 minutes.

## Sampling the Certification Exam:

1. What is the military time for 11:00?
   a. 1100
   b. 2300
   c. 2200
   d. Can't tell

   Answer: _____

2. What is the military time for 12:59 PM?
   a. 16:59
   b. 24:59
   c. 00:59
   d. 12:59

   Answer: _____

3. Mary is taking her amoxicillin every 6 hours. If her last dose was at 1345, when will she take her next dose?
   a. 5:45 PM
   b. 6:45 PM
   c. 7:45 PM
   d. 8:45 PM

   Answer: _____

4. What is 0149 in AM/PM time?
   a. 12:49 PM
   b. 12:49 AM
   c. 1:49 AM
   d. 1:49 PM

   Answer: _____

5. What is 8:30 PM in military time?
   a. 2030
   b. 1830
   c. 1630
   d. 0830

   Answer: _____

## Lesson 8 Content Check

1. What is the military time for 2:30 PM?

   Answer: _____

2. What is the military time for 10:58 PM?

   Answer: _____

3. What is the military time for 12:03 AM?

   Answer: _____

4. What is the military time for 12:20 PM?

   Answer: _____

5. What is the military time for 3:57 AM?

   Answer: _____

6. What is the AM/PM time for 1907?

   Answer: _____

LESSON 8: Time   65

7. What is the AM/PM time for 0355?

   Answer: _____

8. What is the AM/PM time for 2245?

   Answer: _____

9. What is the AM/PM time for 1300?

   Answer: _____

10. What is the AM/PM time for 0145?

    Answer: _____

11. If a patient is given a medication at 9:00 AM and the directions say for them to take it every 8 hours, what time will their next dose be in military time?

    Answer: _____

12. A patient in the hospital has orders to be given meds every 8 hours. If their first dose was given at 8:00 am, what time will their next 3 doses be given in military time?

    Answer: _____

13. A patient is instructed to begin their medication prep for a colonoscopy at 2200 the evening before their procedure. At what time should the patient begin their therapy in standard am/pm time?

    Answer: _____

14. A nurse is reviewing patient charts when she notices that pain medication has been given to a patient at 1100, 1330 and 1700. What times are these in standard am/pm time?

    Answer: _____

15. A patient was told to take her colonoscopy preparation tablets at 0800, 1100, 1400, 1700, and 2100. What times should you tell her so that she can understand them more easily?

    Answer: _____

16. Fill in the following 24-hour chart:

| AM/PM | Military |
|---|---|
| 12:00 AM | |
| 1:00 AM | |
| | 0200 |
| 3:00 AM | |
| | 0400 |
| | 0500 |
| 6:00 AM | |
| | 0700 |
| 8:00 AM | |
| | 0900 |
| 10:00 AM | |
| 11:00 AM | |
| | 1200 |
| 1:00 PM | |
| 2:00 PM | |
| | 1500 |
| 4:00 PM | |
| 5:00 PM | |
| | 1800 |
| 7:00 PM | |
| | 2000 |
| | 2100 |
| 10:00 PM | |
| 11:00 PM | |
| | 0000 |

## Unit 1 Content Review

1. Express the following using Roman numerals.

    a. 13 =

    Answer: _____

    b. 99 =

    Answer: _____

LESSON 8: Time    67

   c. 521 =

   Answer: _____

   d. 1015 =

   Answer: _____

2. Express the following Roman numerals as Arabic numbers.

   a. III =

   Answer: _____

   b. IX =

   Answer: _____

   c. XXVI =

   Answer: _____

   d. MMDLXIV =

   Answer: _____

3. Reduce each fraction to lowest terms.

   a. $\frac{2}{22} =$

   Answer: _____

   b. $\frac{15}{50} =$

   Answer: _____

   c. $\frac{4}{24} =$

   Answer: _____

   d. $\frac{82}{242} =$

   Answer: _____

4. Convert the improper fractions to mixed fractions or whole numbers.

   a. $\frac{19}{3} =$

   Answer: _____

   b. $\frac{45}{6} =$

   Answer: _____

UNIT 1: Basic Mathematics

c. $\dfrac{187}{25} =$

Answer: _____

d. $\dfrac{17}{3} =$

Answer: _____

5. Convert the mixed fractions to improper fractions. Reduce to lowest terms, if possible.

a. $6 \ \& \ \dfrac{2}{5} =$

Answer: _____

b. $1 \ \& \ \dfrac{1}{20} =$

Answer: _____

c. $150 \ \& \ \dfrac{2}{3} =$

Answer: _____

d. $57 \ \& \ \dfrac{11}{13} =$

Answer: _____

6. Multiply or divide the fractions. Reduce to lowest terms, if possible.

a. $\dfrac{1}{4} \times \dfrac{1}{6} =$

Answer: _____

b. $3\dfrac{3}{4} \times 5\dfrac{2}{9} =$

Answer: _____

c. $\dfrac{2}{7} \div 2\dfrac{3}{8} =$

Answer: _____

d. $\dfrac{3/12}{8/16} =$

Answer: _____

7. Convert the fractions to decimal form. Round to the hundredth place, if possible.

a. $\dfrac{2}{10} =$

Answer: _____

b. $\dfrac{4}{20} =$

Answer: _____

c. $9\frac{19}{100} =$

Answer: _____

d. $\frac{5}{1000} =$

Answer: _____

8. Convert the decimals to fractions. Reduce to lowest terms, if possible.

   a. $0.09 =$

   Answer: _____

   b. $0.045 =$

   Answer: _____

   c. $0.004 =$

   Answer: _____

   d. $0.25 =$

   Answer: _____

9. Round the following number to the nearest indicated place value: 1,891.5623

   a. tens place: _____
   b. hundredths place: _____
   c. tenths place: _____
   d. thousandths place: _____

10. Solve the following percent problems. Round to the hundredths place, if possible.

    a. 0.5% of 25

    Answer: _____

    b. 20% of what number is 50?

    Answer: _____

    c. 42% of what number is 90?

    Answer: _____

    d. 5 is what percent of 50?

    Answer: _____

# UNIT 1: Basic Mathematics

11. Write the fractions as a percent. Round to the hundredths place, if possible.

    a. $\frac{7}{10} =$

    Answer: _____

    b. $\frac{1}{6} =$

    Answer: _____

    c. $\frac{5}{7} =$

    Answer: _____

    d. $\frac{8}{9} =$

    Answer: _____

12. Write the ratio's as fractions. Reduce to lowest terms, if possible.

    a. $7 : 4 =$

    Answer: _____

    b. $20 : 2 =$

    Answer: _____

    c. $1 : 20 =$

    Answer: _____

    d. $1 : 150 =$

    Answer: _____

13. Express the fractions as decimals. Round to the hundredths place, if possible.

    a. $\frac{8}{5} =$

    Answer: _____

    b. $\frac{26}{16} =$

    Answer: _____

    c. $\frac{84}{22} =$

    Answer: _____

    d. $\frac{12}{28} =$

    Answer: _____

LESSON 8: Time

14. Solve for X in the proportions. Round to the tenths place, if possible.

   a. $\frac{2}{3} = \frac{20}{X}$

   Answer: _____

   b. $\frac{6}{16} = \frac{X}{32}$

   Answer: _____

   c. $\frac{4}{X} = \frac{12}{9}$

   Answer: _____

   d. $\frac{X}{7.4} = \frac{11}{23}$

   Answer: _____

15. Change the ratio's and decimals to percent's. Round to the thousandths place, if possible.

   a. $5 : 7 =$

   Answer: _____

   b. $4 : 10 =$

   Answer: _____

   c. $0.142 =$

   Answer: _____

   d. $0.05 =$

   Answer: _____

16. Determine whether or not the following are true proportions:

   a. $\frac{37}{273} = \frac{93}{690}$

   Answer: _____

   b. $\frac{12}{240} = \frac{60}{1200}$

   Answer: _____

   c. $\frac{512}{3} = \frac{1536}{9}$

   Answer: _____

   d. $\frac{0.9}{1.7} = \frac{5.2}{9.5}$

   Answer: _____

UNIT 1: Basic Mathematics

17. Convert the following temperatures to either Celsius or Fahrenheit. Round to the tenths place, if possible.

   a. 104°F

   Answer: _____

   b. 39.2°C

   Answer: _____

   c. 18°F

   Answer: _____

   d. -20°C

   Answer: _____

18. Convert the following to either military or standard time.

   a. 1203

   Answer: _____

   b. half past six AM

   Answer: _____

   c. 0032

   Answer: _____

   d. noon

   Answer: _____

# UNIT 2

# Introduction to Units and Problem-Solving Methodology

# LESSON 9

# Systems of Measurement and the Metric System

What makes medical math different from other types of math in the world is the emphasis on the **value** of the number being looked at. This value allows us to assign significance to a given drug dose, medical event, quantity of dosing, etc. In the field of healthcare, the **units** that follow a number are what allow us to attach value, or significance, of any given information.

There are 3 main unit systems used in healthcare. Think of each of these systems as being separate languages – all are attempting to convey the same **value** but use different vocabulary (the units) to get there.

1. _____
   a. Measures two properties (in healthcare):
      i. _____ where the standard of value is _____
      ii. _____ where the standard of value is _____
   b. All other units within this system are variations of the standards.
   c. Most commonly used and understood by healthcare practitioners across the world, but often not understood throughout the USA due to our use of the Imperial System (ex: lbs, inches, etc.). This often makes discussing things with patients confusing as it turns into a language barrier!

2. _____
   a. Typically used to measure the property of _____
   b. Most easily understood by patients due to its use in baking and cooking (ex: tsp., tbsp., oz., etc.)

3. _____
   a. Old, very outdated, and only truly used by pharmacy personnel (ex: dram, grain, minim, etc.)

Regardless of what system is being used, the unit of measurement always (**comes before / goes after**) a number when describing the value of it.

1. Give an example. _____

## Translating Between the Systems

You will need to be able to move around within and between these systems in order to effectively communicate to patients and other members of the healthcare team. There are a few things that will need to be memorized, but others can be understood at a conceptual level.

### The Metric System

Due to the fact that **all units within this system are variations of the standard measurements**, this system is easier to understand at a conceptual level. The metric system makes use of prefixes to adjust the value of the standard, or base, measurement. In healthcare, we only are concerned with two standard measurements: weight, and volume.

Let's discuss a mnemonic that you can use to help you translate units within this system. I call this the KSMM method, and due to it being a mnemonic, it is a shortcut to converting between metric units.

|  | **K** | **S** | **M** | **M** |
|---|---|---|---|---|
| **Acronym Mnemonic:** |  |  |  |  |
| **Prefix:** |  |  |  |  |
| **Prefix Abbreviation:** |  |  |  |  |
| **Meaning:** |  |  |  |  |

How to use the mnemonic:

1. Ensure that you are translating weight to weight, or volume to volume by checking to see if the standard (base) unit is the same in both the starting and ending point. It is VERY IMPORTANT to note that you CANNOT use this mnemonic to convert weight to volume, or vice versa.

   > **Example:** 200 mg -> mcg

   *Can I convert m**g** into mc**g**? YES – this conversion is weight to weight due to the base being g for grams

2. Identify the unit that you need to convert – as if you were on a road trip, ask yourself where do I begin? Place a big dot underneath the letter of the acronym that is associated with where you start.

           200 mg -> mcg     K S Ṃ M     *Starting with milligrams

## LESSON 9: Systems of Measurement and the Metric System

3. Draw one curved arrow PER letter beginning at the starting dot and ending at the letter of the acronym indicating the unit you are converting to. If the desired unit is 2 letters away, draw 2 curved arrows. If the desired unit is 3 letters away, draw 3 arrows. Note that you will have at most 3 arrows in any given conversion.

    200 mg -> mcg     K S Ṃ M

4. Rewrite out the number to be converted without units. Indicate where the decimal place is if there is not a decimal already shown. All whole numbers have an implied decimal at the end of the number.

5. Move the decimal in the number **3 times per letter** in the **same direction** that you drew the arrow in.

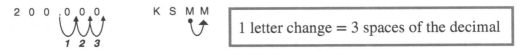

6. Fill in the holes with zeros, if there are any, and be sure to indicate leading zeros where there are some.

    2 0 0, 0 0 0.

7. Rewrite the number with the new unit. Watch your zero's and add commas where necessary.

    > 200,000 mcg

    > The KSMM method eliminates the need to memorize the following conversions within the metric system:
    >
    > | | | |
    > |---|---|---|
    > | 1 kg = 1,000 g | 1 kg = 1,000,000 mg | 1 kg = 1,000,000,000 mcg |
    > | 1 g = 1,000 mg | 1 g = 1,000,000 mcg | 1 g = 0.001 kg |
    > | 1 mg = 1,000 mcg | 1 mg = 0.000001 kg | 1 mcg = 0.000000001 kg |
    > | 1 mg = 0.001g | 1 mcg = 0.000001 g | 1 mcg = 0.001 mg |

**STOP AND PRACTICE:** Prove that this shortcut works by converting the following numbers within the metric system by using the short cut and then comparing the answers to the chart above.

UNIT 2: Introduction to Units and Problem-Solving Methodology

1.  1 kg -> g          K S M M

    Answer: _____

2.  1 g -> mg          K S M M

    Answer: _____

3.  1 mg -> mcg        K S M M

    Answer: _____

4.  1 mcg -> mg        K S M M

    Answer: _____

5.  1 mg -> g          K S M M

    Answer: _____

6.  1 g -> kg          K S M M

    Answer: _____

## PAUSE AND REVIEW:

1. To start a conversion using the KSMM method, you first need to know _____ _____, then you determine _____
2. How many "steps" are in between in each letter in the KSMM method? _____
   a. The "steps" between each letter represents _____ moving _____ times in (**the same / opposite**) direction as the letters.
3. **T / F:** The KSMM method is a shortcut to conversions within the metric system.
4. **T / F:** It is ok to use the KSMM method to convert units within the metric system as long as they are expressing the same thing (ex: weight-to-weight, or volume-to-volume)
   a. Why? _____
   _____

**LESSON 9:** Systems of Measurement and the Metric System

5. **T / F:** Always use a leading zero and NEVER use a trailing zero.
    a. Why? _____
    _____
    _____

6. **T / F:** All whole numbers have *implied* decimals at the end.
    a. Give an example: _____

**STOP AND PRACTICE:** Convert the following numbers within the metric system.

1. 1.29 g -> mg        K S M M

    Answer: _____

2. 24.7 mcg -> mg      K S M M

    Answer: _____

3. 1420 mg -> kg       K S M M

    Answer: _____

4. 495 mL -> L         K S M M

    Answer: _____

5. 3.781 kg -> mcg     K S M M

    Answer: _____

## Sampling the Certification Exam:

1. Convert 12 L -> mL
    a. 1.2 mL
    b. 0.012mL
    c. 12,000 mL
    d. 120 mL

    Answer: _____

# UNIT 2: Introduction to Units and Problem-Solving Methodology

2. Convert 530 mcg -> g
   a. 0.530 g
   b. 0.0530 g
   c. 0.00530 g
   d. 0.000530 g

   Answer: _____

3. What is the prefix for 1/1000th?
   a. kilo-
   b. milli-
   c. micro-
   d. N/A – this is the standard

   Answer: _____

4. A full medication stock bottle weighs 2,678 g. How many kg is this?
   a. 2.678 kg
   b. 26.78 kg
   c. 2,678,000 kg
   d. 267.8 kg

   Answer: _____

5. A newborn baby weighs 3.588 kg. How many g does he weigh?
   a. 0.003588 g
   b. 0.3588 g
   c. 358.8 g
   d. 3,588 g

   Answer: _____

## Lesson 9 Content Check

1. Convert 413 mg -> g

   Answer: _____

**LESSON 9:** Systems of Measurement and the Metric System

2. Convert 5 mL -> L

   Answer: _____

3. Convert 5.3 kg -> mg

   Answer: _____

4. Convert 0.349 g -> mg

   Answer: _____

5. Convert 1.5 g -> mcg

   Answer: _____

6. Convert 0.083 g -> mg

   Answer: _____

7. Convert 2749 mcg -> kg

   Answer: _____

8. Convert 2.1 g -> mg

   Answer: _____

9. Convert 948 mL -> L

   Answer: _____

10. Convert 0.02 L -> mL

    Answer: _____

11. Convert 1402 mL -> L

    Answer: _____

12. Convert 10 mL -> L

    Answer: _____

UNIT 2: Introduction to Units and Problem-Solving Methodology

13. Convert 120,594 mL -> L

    Answer: _____

14. Convert 0.0043 g -> mcg

    Answer: _____

15. Convert: 4,975 mcg to g

    Answer: _____

16. Convert 795 kg -> g

    Answer: _____

17. Convert 94 L -> mcL

    Answer: _____

18. Convert 3745 mcg to milligrams.

    Answer: _____

19. Convert the following: 50 mcg -> g

    Answer: _____

20. Convert the following: 0.12 kg -> mg

    Answer: _____

# LESSON 10

# The Household and Apothecary Measuring System

## The Household System

This is the system most Americans know due to _____!

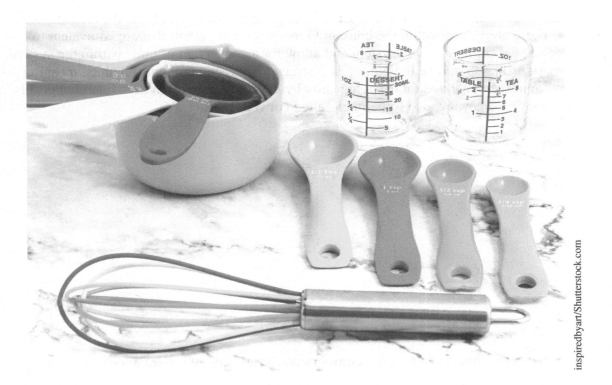

Fill in the chart of the units in this system below:

| Unit | | | | | | | |
|---|---|---|---|---|---|---|---|
| Abbreviation | | | | | | | |

This chart, when read from left to right, puts these units in (**increasing / decreasing**) order.

# UNIT 2: Introduction to Units and Problem-Solving Methodology

## Side note regarding ounces:

1. What is the difference between a fl. oz. and oz.? _____
   _____
   _____

   a. What unit within the metric system is associated with each?
      fl. oz. = _____
      oz. = _____

2. How do you know if a problem is talking about a volume or a weight, regarding ounces? _____
   _____
   _____

These units can be combined with numbers to represent a value (just like in the metric system). There are values within the household measurement system that are equivalent to (=) other values within the same system and allow for a conversion of units within the system. These relationships should be thought of as marriages between numbers and units, meaning that they cannot be separated (at least, logically by math), and are often used interchangeably with one another. They are also sometimes called **conversion factors**.

$$= $$
$$= $$
$$= $$
$$= $$
$$= $$
$$= $$

*Tips and Tricks:*

1. What is the sequence of all the **numbers** on the left side of the equation? _____

2. Regarding the **units** on the left side of the equation, the values (**increase / decrease**) As you move from top to bottom.

3. What is the sequence of all the **numbers** on the right side of the equation? _____

4. What do you notice about all of the **units** on the right side of the equation? _____
   _____
   _____

LESSON 10: The Household and Apothecary Measuring System

5. What can you do to make your memorization easier? _____
_____

## The Apothecary System:

What is the most important (and only) apothecary value that needs to be recognized?

= 

1. Do you need to memorize this value? Why/why not? _____
_____

Concerning other apothecary abbreviations that may come up:

1. Which unit of the Apothecary system is outdated and not used? _____
    a. What is the abbreviation for this unit? _____
2. Which unit of the Apothecary system is used to express how large a prescription vial is? _____
    a. What is the abbreviation for this unit? _____
3. What are the common sizes of prescription vials? _____
_____

## Other Important Units of Measurement:

The only other values that need to be memorized in order to be able to convert between any unit systems are:

1. 1 tsp = _____
2. 1 kg = _____
3. (Insulin) 1 mL = _____ *

> ***Be careful with insulin**. The drug label you are working with should be VERY clear on what the relationship is, but more information regarding insulin will be discussed later in this text.

Fill in the following chart for the other units that may come up in pharmacy math problems

| Unit | Units | Milliequivalents | Drops |
|---|---|---|---|
| Abbreviation | | | |
| Conversion | | | |
| Example | | | |

# UNIT 2: Introduction to Units and Problem-Solving Methodology

**STOP AND PRACTICE:** Spell out the following numbers and units:

1. 3 gal: _____
2. ½ gr: _____
3. 6 U: _____
4. 16 mEq: _____
5. 17 gtt: _____

## Mathematical Expression of Values:

1. What 3 ways can these relationships be expressed mathematically? Give examples


2. In a fraction, the "/" symbol can indicate _____ or a _____ of two values.

3. **T / F:** Replacing an "=" sign in a relationship between two conversion factors with a "/" sign changes its relationship value.

4. **T / F:** Flipping a fraction over changes its relationship/value.

   a. Explain: _____
   _____
   _____

   b. When/why would you do this? _____
   _____
   _____

**STOP AND PRACTICE:** Rewrite the conversion factors learned in this lesson below as different mathematical expressions:

| Equal Sign | Fraction | Ratio |
|---|---|---|
| 1 tsp = 5 mL | | |
| | 1 kg / 2.2 lb | |
| | | 1 mL : 100 units (of U-100 insulin) |
| 1 gr = 60 mg | | |
| | 1 tbsp / 3 tsp | |
| | | 1 oz : 2 tbsp |
| 1 cup = 8 oz | | |
| | 1 pt / 2 cup | |

**LESSON 10:** The Household and Apothecary Measuring System   87

| Equal Sign | Fraction | Ratio |
|---|---|---|
|  |  | 1 qt : 2 pt |
| 1 gal = 4 qt |  |  |

## Applying Concepts:

1. Match the household unit to its position in the following picture to help you get a better mental image of the size of these household units:

   a. cup

   b. tbsp.

   c. tsp

   d. oz

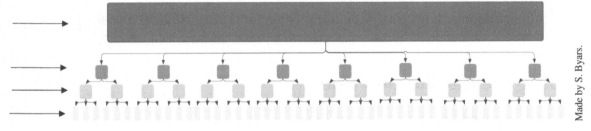

2. Define the household unit in its position in the following picture to help you get a better mental image of the size of household units:

   a. _____

   b. _____

   c. _____

   d. _____

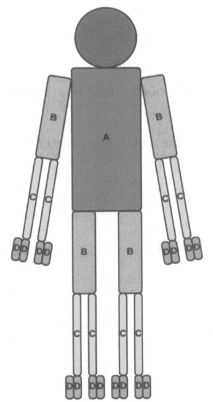

3. Here is another image that you may have seen before:

Image source: https://www.pinterest.com/pin/316729786265020064/

## Sampling the Certification Exam:

1. How many milliequivalents are in 3 mg?
   a. 30 mEq
   b. 12.5 mEq
   c. 15 mEq
   d. none - there is no conversion factor for milliequivalents

   Answer: _____

2. One tablespoon is equal to ___ teaspoons.
   a. 1
   b. 2
   c. 3
   d. 4

   Answer: _____

3. One gallon is equal to __ quarts.
   a. 1
   b. 2
   c. 3
   d. 4

   Answer: _____

4. One cup is equal to __ ounces.
   a. 2
   b. 4
   c. 6
   d. 8

   Answer: _____

5. One pint is equal to __ cups.
   a. 1
   b. 2
   c. 3
   d. 4

   Answer: _____

## Lesson 10 Content Check

1. What is the order of digits in the Household unit sequence?

   Answer: _____

2. What is the order of units in the Household unit sequence in order from largest to smallest?

   Answer: _____

3. How many pounds are in 1 kg?

   Answer: _____

4. How many pts are in 1 qt?

Answer: _____

5. How many tablespoonfuls are in 1 ounce?

Answer: _____

6. How many quarts are in 1 gallon?

Answer: _____

7. How many cups are in 1 pint?

Answer: _____

8. How many teaspoonfuls are in 1 tbsp?

Answer: _____

9. How many ounces are in 1 cup?

Answer: _____

10. How many mL are in 1 tsp?

Answer: _____

11. Amoxicillin suspension would be measured in fluid ounces or solid ounces?

Answer: _____

12. A newborn baby weighs seven pounds and eight ounces. How can this be represented using only numbers and abbreviations?

Answer: _____

13. An old textbook refers to something as being twelve minims. How can this be represented using only numbers and abbreviations?

Answer: _____

14. A prescriber gives a verbal order for four thousand units of a drug. How can this be represented using only numbers and abbreviations?

Answer: _____

**LESSON 10:** The Household and Apothecary Measuring System

15. The conversion between milliliters and teaspoons is a conversion of which two measuring systems?

    Answer: _____

16. The unit "gr" is from which measuring system?

    Answer: _____

17. The unit "tbsp." is from which measuring system?

    Answer: _____

18. The unit "mEq" stands for:

    Answer: _____

19. The unit "gtt" stands for:

    Answer: _____

20. The unit "dr" stands for:

    Answer: _____

# LESSON 11

# Understanding Math Word Problems (Math Literacy)

Most people who say they are "bad at math" just struggle with reading and understanding a word problem. In order to combat this, there are 4 simple steps you can follow:

Step 1) **Identify the type of problem.**

Use key words within the problem to determine what you are solving for (See appendix for further discussion). Focus on words within the problem that make it different from other problems and highlight those words.

*Tips and Tricks:* It is ideal to always read a question 3 times:

1st read – Prepares your brain for the general idea of the problem.

2nd read – Helps you focus more and identify key words that will help you understand how you will solve the problem.

3rd read – This should occur AFTER you solve a problem. This is the point where you might catch any errors you made in haste – does your answer make sense?

Step 2) **Determine the method you will use to solve the problem.**

Most of all medical math problems can be solved using either the ratio/proportion method or the dimensional analysis method (explained in the next few lessons. and the method you choose is dictated largely by personal choice, but then lends itself naturally to a structured set up.

Other problems require memorized formulas/methodical approaches (i.e., alligations. and will be discussed as we move through the text. There are only a few types of these problems within the math you will encounter in the world of pharmacy.

Step 3) **Set up the problem**.

Follow the steps to setting up the problem according to the method/formula/ approach you chose to go with. Deviations from those steps are what cause errors – do *not* take shortcuts!

# UNIT 2: Introduction to Units and Problem-Solving Methodology

Step 4) **Solve the problem.**

Follow the steps to solve the problem according to the method/formula/approach you chose to go with. Let the math and your previous logic guide your process!

*Tips and Tricks:*

1. In order to use the ratio/proportion and dimensional analysis method, it helps to understand how to identify the "knowns" and "unknowns" in any given problem.

   **A "known" in medical math will always be:**

   -What you physically HAVE in your hands to work with (i.e., what is in stock)

   -A drug's strength or concentration

   -A conversion factor (ex: 1 kg = 2.2 lbs)

   -Any other relationship between values that cannot be separated (often, given in the problem itself)

   **An "unknown" in medical math will always be:**

   -What a prescriber orders/prescribes

   -What is "needed" in the problem (ex: the dose a patient needs)

   No matter what – they BOTH have to be represented as a fraction in order to make either method work!

2. If the question is asking for a unit that does not match the units in the known variable (i.e., it wants you to convert), HIGHLIGHT it so that you don't forget to convert that unit!

   a. In this case, it's best to wait to convert or round until the END of a problem if possible.

   b. Never convert the known – always convert the unknown if conversion is necessary!

3. **T / F:** Saying something is "100 mg per mL" is the same thing as the mathematical expression, 100 mg/1 mL

   c   Therefore, we can assume that "per" indicates what? _____

4. **T / F:** 5 mL per teaspoon is written like this: 5 mL/tsp.

   d   How many teaspoons are implied? _____

5. **T / F:** A 10 mg tablet can be written mathematically as 10 mg/1 tablet without changing the value.

**LESSON 11:** Understanding Math Word Problems (Math Literacy)    **95**

**STOP AND PRACTICE:** Identify different parts of the following problems.

1. Progesterone is available as 10 mg/1 mL solution in your pharmacy. A prescription calls for a dose of 200 mg of progesterone. How many milliliters are needed for that dose?

    a. What is the known? _____

        i. How do you know this? _____

    b. What is the unknown? _____

        i. How do you know this? _____

    c. What unit should your answer be in? _____

2. Ciprofloxacin is in stock at your pharmacy as a 500 mg tablet. How many tablets are needed to give a dose of 1,250 mg?

    a. What is the known? _____

        i. How do you know this? _____

    b. What is the unknown? _____

        i. How do you know this? _____

    c. What unit should your answer be in? _____

3. Terbinafine comes from the manufacturer as a solution with a concentration of 75 mg/mL. How many milliliters are needed to prepare a foot soak that contains 325 mg?

    a. What is the known? _____

        i. How do you know this? _____

    b. What is the unknown? _____

        ii. How do you know this? _____

    c. What unit should your answer be in? _____

4. A dentist orders a one-time dose of 2 grams of amoxicillin for a patient. You have a 500 mg capsule available to use in your inventory. How many capsules will the patient need for their dose?

    a. What is the known? _____

        i. How do you know this? _____

    b. What is the unknown? _____

        ii. How do you know this? _____

    c. What unit should your answer be in? _____

**UNIT 2:** Introduction to Units and Problem-Solving Methodology

5. A canine patient needs 100 mg of zonisamide to help control their seizures. If zonisamide is available as a 1 gram/10 mL suspension, how many milliliters would contain the appropriate dose?

   a. What is the known? _____

      i. How do you know this? _____

   b. What is the unknown? _____

      ii. How do you know this? _____

   c. What unit should your answer be in? _____

# LESSON 12
# The Ratio/Proportion Method

The only difference between the proportion problems we've solved previously and those in all future problems is the presence of units. We will now discuss how to use the ratio/proportion method with the addition of units to solve for almost any medical math problem.

Let's use the following problem to discuss this in more depth:

> **Example:** A patient needs 250 mg. On the shelf, the pharmacy stocks a 500 mg tablet. How many tablets would it take to get the patient one dose of medication?

Step 1) Identify and write out the "known" as a fraction.

$$\frac{500 \text{ mg}}{1 \text{ tab}}$$

Step 2) Draw an equal sign to the right of the known, and a dividing line for a new fraction. This is the set up for the "unknown" that you are solving for.

$$\frac{500 \text{ mg}}{1 \text{ tab}} = \frac{\phantom{xx}}{\phantom{xx}}$$

Step 3) Rewrite the units exactly as they appear – the numerator and denominator will stay the same. *In order for this method to work, the units have to line up in exactly this order.*

$$\frac{500 \text{ mg}}{1 \text{ tab}} = \frac{\text{mg}}{\text{tab}}$$

Step 4) Using the units as a guide, plug in the "unknown" where it matches. The last place value should be assigned a generic variable. In math "*X*" is used most frequently.

$$\frac{500 \text{ mg}}{1 \text{ tab}} = \frac{250 \text{ mg}}{X \text{ tab}}$$

Step 5) Using the numbers only and the principles of cross multiplication, solve for *X*.

$$\frac{500 \text{ mg}}{1 \text{ tab}} = \frac{250 \text{ mg}}{X \text{ tab}} \qquad \text{Where } X = \frac{250 \times 1}{500} = \frac{1}{2}$$

Step 6) Assign the unit associated with the unknown number within the proportion.

½ tab

Note that it does not matter how the known (step 1) is initially set up as long as steps 2-6 are followed.

$$\text{Proof}: \frac{1 \text{ tab}}{500 \text{ mg}} = \frac{X \text{ tab}}{250 \text{ mg}} \text{ Where } X = \frac{250 \times 1}{500} = \frac{1}{2} \text{ tab}$$

**STOP AND PRACTICE:** Calculate the following dosages by following the steps above:

1. A patient needs 10 mg of medication. You have the following on hand: 25 mg/5 mL. How many mL are needed to get the patient's dose?

    Answer: _____

2. A patient needs 1.5 g of medication. You have the medication on hand in a strength of 3g/10 mL. How many mL are needed to get the patient's dose?

    Answer: _____

3. A patient needs 25 mg of medication. You have 50 mg tablets on hand. How many tablets are needed to get the patient's dose?

    Answer: _____

4. A patient needs 75 mg of medication. You have 25 mg capsules on hand. How many capsules are needed to get the patient's dose?

    Answer: _____

If you get to step 4 and you pause because the units given in the problem **don't match** what you have set up, you will need to convert using the KSMM method, or by using the other conversion factors discussed above.

> **Example:** A patient needs 0.3 g of medication that is available in the pharmacy as a 750 mg/ 5 mL suspension. How many mL of the suspension would provide the patient with their appropriate dose?

Step 1) The known in this problem is: $\frac{750 \text{ mg}}{5 \text{ mL}}$

Step 2) $\frac{750 \text{ mg}}{5 \text{ mL}} = \frac{}{}$

**LESSON 12:** The Ratio/Proportion Method

Step 3) $\dfrac{750\,mg}{5\,mL} = \dfrac{mg}{mL}$

Step 4) $\dfrac{750\,mg}{5\,mL} = \dfrac{mg}{mL}$

> *The only other piece of information given in the problem is 0.3 g, which doesn't fit anywhere given how we have the problem set up to solve.

*PAUSE!* We need 0.3 g to be converted either to mg, or mL to be able to fit into the structure we have set up. Since the unit g indicates a weight, we can only convert to mg (another measurement that indicates weight). Remember: the KSMM method uses conversions between units of the same type (weight to weight, or volume to volume)

$0.3\,g \rightarrow mg = 300\,mg$

Now that the unit matches a unit in the set up, we can continue working through the problem as before with the following changes:

$\dfrac{750\,mg}{5\,mL} = \dfrac{300\,mg}{X\,mL}$

Step 5) $\dfrac{750\,mg}{5\,mL} = \dfrac{300\,mg}{X\,mL}$   Where $X = \dfrac{300 \times 5}{750} = 2$

Step 6) 2 mL

**STOP AND PRACTICE:** Calculate the following dosages:

1. A patient needs 250 mg of medication. On hand, there are 0.5 g tablets. How many tablets are needed to obtain the patient's dose?

   Answer: _____

2. A patient needs 0.5 mg of medication. You have 100 mcg tablets on hand. How many tablets are needed to obtain the correct dose?

   Answer: _____

3. A patient needs 50 mcg of a medication. You have 0.1 mg tablets on hand. How many tablets are needed for the patient's dose?

   Answer: _____

4. A patient needs 1,500 mg of medication but you only have a suspension with a strength of 5 g/ 500 mL on hand. How many mL will it take to give the patient their dose of medication?

Answer: _____

Some proportion questions will not define the known very well – it will come from a memorized unit conversion mentioned in previous lessons.

**Example:** Convert 84 lb -> kg

Step 1) First ask yourself, what is the "known" conversion between the units?

1 kg = 2.2 lb

Step 2) Then, set up a proportion and solve.

$$\frac{1\,kg}{2.2\,lb} = \frac{X\,kg}{84\,lb} \text{ Where } X = \frac{84 \times 1}{2.2} = \boxed{38.18\,kg}$$

**STOP AND PRACTICE:** Convert the following using the household or apothecary system:

1. 9 tsp. -> tbsp.

Answer: _____

2. 30 tbsp. -> oz.

Answer: _____

3. 4 cup -> oz.

Answer: _____

4. 16 pt. -> qt.

Answer: _____

5. 4 gal -> qt.

Answer: _____

## Sampling the Certification Exam:

1. A patient weighs 67 kg. How many pounds does he weigh?

    a. 13.4 lb

    b. 30.45 lb

    c. 147.4 lb

    d. 335 lb

Answer: _____

2. A patient is ordered to take 2 tablespoonfuls of medication. How many teaspoonfuls does this equal?

   a. 2/3 tsp.
   b. 4 tsp.
   c. 6 tsp.
   d. 8 tsp.

   Answer: _____

3. A patient has a prescription order for 30 mg of a medication. The pharmacy has a partial container of 7.5 mg tablets in stock. How many tablets will the patient need for their prescribed dose?

   a. 4 tablets
   b. 4.5 tablets
   c. 5 tablets
   d. 7 tablets

   Answer: _____

4. A prescriber ordered a dose of 100 mg minocycline oral solution. The medication is available in a 0.5 g/5 mL oral solution. How many milliliters are needed to provide the ordered dose?

   a. 0.5 mL
   b. 1 mL
   c. 2 mL
   d. 5 mL

   Answer: _____

5. A prescriber ordered a dose of 0.6 g of a medication that is available as a 500 mg/10 mL solution. How many milliliters are needed to provide the ordered dose?

   a. 7.5 mL
   b. 10 mL
   c. 12 mL
   d. 15 mL

   Answer: _____

## Lesson 12 Content Check

1. Convert 6 oz. -> cups and express as a fraction.

    Answer: _____

2. Convert 45 tsp -> tbsp.

    Answer: _____

3. Convert 13 gal -> qt.

    Answer: _____

4. How many 100 mcg tablets are needed to complete an order for 500 mcg worth of medicine?

    Answer: _____

5. How many 0.1 mg tablets are needed to complete an order for 0.5 mg worth of medication?

    Answer: _____

6. How many mL of a 3 g/10 mL solution is needed to get a dose of 1.5 g?

    Answer: _____

7. How many grams of a drug with a strength of 5 g/500 mL will 0.2 L yield?

    Answer: _____

8. How many 0.5 mg tablets would be needed for a 250 mcg dose?

    Answer: _____

9. Convert 18 oz. -> cup

    Answer: _____

10. Convert 14 pt. -> qt.

    Answer: _____

11. Solve the following: 4 cup -> pt.

    Answer: _____

12. Convert 2.3 gal -> qt.

    Answer: _____

13. Convert 3 tbsp. -> tsp.

    Answer: _____

14. Convert the following: 96 qt -> gal

    Answer: _____

15. Convert the following: 4.2 gal -> qt

    Answer: _____

16. Convert the following: 25 mL -> gtt (where 1 mL = 20 gtt)

    Answer: _____

17. Convert the following: 5 tbsp -> oz

    Answer: _____

18. Convert the following: 152 lb -> kg

    Answer: _____

19. Convert the following: 300 mg -> gr (where 1 gr = 65 mg)

    Answer: _____

20. A prescriber ordered a dose of 60 mg of doxycycline. The medication is available as a 20 mg/2 mL solution. How many milliliters are needed to provide the ordered dose?

    Answer: _____

# LESSON 13

# Multistep Ratio/Proportion Problems

Medical math problems are not always straightforward, and often require some critical thinking to be able to translate what is needed using what you have available. A good analogy is to compare this process to using a road map or GPS while driving to an unfamiliar location. To start, you must know/input two pieces of information – where you are starting from (for this methodology, the "known"), and where you want to end up (for this methodology, the "unknown"), then your map/GPS gives you step-by-step instructions on how to get there. In these instructions, you cannot skip steps, or turns, if you want to end up in the right place. The same is true of math problems in that you have to make the right "turns" to solve the problem.

In terms of medical math problems, a good example is to use a conversion within the household system.

> **Example:** Convert 32 tablespoons to cups.

Step 1. Outline your road map. Do this by reviewing the information in the problem and determine where you are starting from, and where you need to go. If there is a direct link (an existing conversion discussed so far in this text) between these values, then no real map would be needed. If there is not a direct link (no conversions between the values discussed so far in this text), you must determine how you can get from one unit to the next without skipping steps in between.

1. Convert tablespoons to ounces
2. Convert ounces to cups

> We must do this because there is not an established link (a conversion) between tablespoons and cups discussed in this text!

Step 2. Follow the road map and use the principles of the ratio/proportion method to solve the problem.

1. $\dfrac{2 \text{ tbsp}}{1 \text{ oz}} = \dfrac{32 \text{ tbsp}}{X \text{ oz}}$ where $X = 16$ oz.

2. $\dfrac{8 \text{ oz}}{1 \text{ cup}} = \dfrac{16 \text{ oz}}{X \text{ cup}}$ where $X = \boxed{2 \text{ cup}}$

**STOP AND PRACTICE:** Convert the following household units. First, outline your road map, then convert using proportions:

1. 20 tsp. -> oz.

    a. Road Map:

       1.
       
       2.

    b. Work:

       1.
       
       2.

    Answer: _____

2. 8 cups -> tbsp.

    a. Road Map:

       1.
       
       2.

    b. Work:

       1.
       
       2.

    Answer: _____

3. 20 pt -> tbsp

    a. Road Map:

       1.
       
       2.
       
       3.

    b. Work:

       1.
       
       2.
       
       3.

    Answer: _____

4. 5 gal -> oz.

   a. Road Map:
      1.
      2.
      3.
      4.

   b. Work:
      1.
      2.
      3.
      4.

   Answer: _____

5. 19.4 qt. -> tsp.

   a. Road Map:
      1.
      2.
      3.
      4.
      5.

   b. Work:
      1.
      2.
      3.
      4.
      5.

   Answer: _____

# UNIT 2: Introduction to Units and Problem-Solving Methodology

## Sampling the Certification Exam:

1. Convert 2.3 gallons to pints.
   a. 9.2 pt.
   b. 18.4 pt.
   c. 24.3 pt.
   d. 27.2 pt.

   Answer: _____

2. Convert 50 ounces to pints.
   a. 3.125 pt.
   b. 6.25 pt.
   c. 25 pt.
   d. 800 pt.

   Answer: _____

3. Convert 26 cups to quarts.
   a. 3.25 qt.
   b. 6.5 qt.
   c. 13 qt.
   d. 52 qt.

   Answer: _____

4. Convert 3.4 pounds to grams.
   a. 1.55 g
   b. 1,545 g
   c. 7.48 g
   d. 7,480 g

   Answer: _____

5. Convert 13,928 grams to pounds.
   a. 6.33 lb
   b. 10.92 lb
   c. 24.41 lb
   d. 30.64 lb

   Answer: _____

## Lesson 13 Content Check

1. A full medication stock bottle weighs 2,678 g. How many lbs is this?

    Answer: _____

2. Convert 89 tbsp. -> pt.

    Answer: _____

3. Convert 6.8 gal -> tsp

    Answer: _____

4. Convert 2 pt. -> oz.

    Answer: _____

5. Convert 39 g -> lb

    Answer: _____

6. Convert 51 cups -> gal

    Answer: _____

7. Convert ½ oz -> tsp

    Answer: _____

8. Convert 5,783 g -> lb

    Answer: _____

9. Convert 3 pt -> mL

    Answer: _____

10. Convert 4 oz -> mL

    Answer: _____

11. Convert 3.1 gallons to milliliters.

    Answer: _____

12. Convert 97 teaspoons to cups.

    Answer: _____

13. Convert 102 tablespoons to gallons.

    Answer: _____

14. Convert 1,802 teaspoons to pints.

    Answer: _____

15. Convert 16 gallons to cups.

    Answer: _____

16. Convert 3 quarts to liters.

    Answer: _____

17. Convert 52.3 pounds to milligrams.

    Answer: _____

18. Convert 1 cup to milliliters.

    Answer: _____

19. Convert 1 ounce to milliliters.

    Answer: _____

20. Convert 1 tablespoon to milliliters.

    Answer: _____

# LESSON 14

# The Dimensional Analysis Method

Dimensional analysis is a method that is used to convert one unit to another by means of multiplication of fractions. The fractions themselves, like the ones used in ratio-proportion, are often called conversion factors.

Let's use the following problem to discuss this in more depth:

> **Example:** A patient is prescribed 13 mL of an antibiotic suspension. How many teaspoons should the pharmacist tell the patient to take?

Step 1) Start by writing out the unknown, with units. If it is not already a fraction, make it one by putting a line underneath and dividing by 1. This does not change the value of the unit because anything divided by 1 is itself!

$$\frac{13 \text{ mL}}{1}$$

Step 2) Draw a multiplication sign to the right, and a dividing line for a new fraction.

$$\frac{13 \text{ mL}}{1} \times -$$

Step 3) Write out the new units ONLY on the new fraction so that the **numerator is the same as the denominator, or vice versa.** This allows us to cancel certain units in order to leave us with the unit we want or are trying to convert to.

$$\frac{13 \text{ mL}}{1} \times \frac{\text{tsp.}}{\text{mL}}$$

> *mL divided by mL = 1, and anything leftover times 1 will be itself

Step 4) (*If needed*) Repeat steps 2 and 3 until you are in the units that you want/are trying to convert to. When you arrive at that unit, and all others have been canceled, put an equal sign to the right of the last fraction.

$$\frac{13 \cancel{\text{mL}}}{1} \times \frac{\text{tsp.}}{\cancel{\text{mL}}} =$$

Step 5) Plug in your knowns using the units already written out to help guide you on where the numbers go.

$$\frac{13\,\text{mL}}{1} \times \frac{1\,\text{tsp.}}{5\,\text{mL}} =$$

> *We know that 1 teaspoon = 5 mL, and from step 3, we know that teaspoons goes on top and mL goes on the bottom, therefore the number 1 goes on top (teaspoons) and the number 5 goes on bottom (mL)

Step 6) Multiply all numbers across the top. Multiply all numbers across the bottom. Divide the resulting products.

$$\frac{13\,\text{mL}}{1} \times \frac{1\,\text{tsp.}}{5\,\text{mL}} = \frac{13 \times 1}{1 \times 5} = \frac{13}{5} = 2.6$$

Step 7) Write the units you determined from Step 4 next to the number.

> 2.6 tsp

**STOP AND PRACTICE:** Calculate the following dosages using the dimensional analysis method:

1. A patient needs 10 mg of medication. You have the following on hand: 25 mg/5 mL. How many mL are needed to get the patient's dose?

   Answer: _____

2. A patient needs 1.5 g of medication. You have the medication on hand in a strength of 3 g/10 mL. How many mL are needed to get the patient's dose?

   Answer: _____

3. A patient needs 25 mg of medication. You have 50 mg tablets on hand. How many tablets are needed to get the patient's dose?

   Answer: _____

4. A patient needs 75 mg of medication. You have 25 mg capsules on hand. How many capsules are needed to get the patient's dose?

   Answer: _____

## Multistep Dimensional Analysis Problems

The same logic of solving multistep ratio/proportion problems applies to solving multistep dimensional analysis problems with the exception that using dimensional analysis allows you to solve any given problem using one equation. Let's look at the multistep example from the ratio/proportion section and apply the dimensional analysis method.

## LESSON 14: The Dimensional Analysis Method

*Tips and Tricks:*

1. As stated above in steps 2-3, write out your units of each fraction/conversion factor FIRST before you put ANY numbers in regardless of how many fractions you have. Cancel them out (top to bottom or vice versa) as you move along. This prevents you from skipping any steps.

    a. To help, ask yourself – "Is this the unit I need to be in?" or "Is this the unit the question is asking for?" or "Is this the unit I need to stop converting to?" These questions can guide you on your journey through the problem.

2. Understand that each fraction/conversion factor after the first one is a "known" – either one you have memorized (ex: 1 kg = 2.2 lb), or one that is given in the problem. **The orientation of the units is really what is the key in dimensional analysis.**

> **Example:** Convert 32 tablespoons to cups.

Using just the units: $\dfrac{32 \text{ tbsp.}}{1} \times \dfrac{\cancel{oz.}}{\cancel{tbsp.}} \times \dfrac{\boxed{cup}}{\cancel{oz.}} =$

Now with the numbers: $\dfrac{32 \text{ tbsp.}}{1} \times \dfrac{1 \text{ oz.}}{2 \text{ tbsp.}} \times \dfrac{1 \text{ cup}}{8 \text{ oz.}} = \dfrac{32 \times 1 \times 1}{1 \times 2 \times 8} = \dfrac{32}{16} = \boxed{2 \text{ cup}}$

**STOP AND PRACTICE:** Use dimensional analysis to solve the same problems from the stop and practice section at the end of the previous lesson. Compare your answers and ensure that they match.

1. 20 tsp. -> oz.

    Answer: _____

2. 8 cup -> tbsp.

    Answer: _____

3. 20 pt. -> tbsp.

    Answer: _____

4. 5 gal -> oz.

    Answer: _____

5. 19.4 qt. -> tsp.

    Answer: _____

## Sampling the Certification Exam:

1. Which method of solving problems requires the units to cancel by matching numerator to denominator?

    a. Ratio/proportion
    b. Dimensional analysis
    c. KSMM
    d. Formula

    Answer: _____

2. Dimensional analysis relies on the mathematical principle that anything multiplied by one is equal to

    a. Zero
    b. Itself
    c. One
    d. N/A

    Answer: _____

3. Dimensional analysis relies on the mathematical principle that a unit divided by itself equals to

    a. Zero
    b. Itself
    c. One
    d. N/A

    Answer: _____

4. The following equation would yield what unit of measurement as an answer?

    $$\frac{5,000\,U}{1} \times \frac{5\,mL}{10,000\,U} \times \frac{20\,gtt}{1\,mL} =$$

    a. gtt
    b. mL
    c. U
    d. None of the above.

    Answer: _____

5. Which of the following represents converting 3 gallons to ounces using the dimensional analysis method?

   a. $\dfrac{3\,\text{gal}}{1} \times \dfrac{4\,\text{qt}}{1\,\text{gal}} \times \dfrac{2\,\text{pt}}{1\,\text{qt}} \times \dfrac{2\,\text{cup}}{1\,\text{pt}} \times \dfrac{8\,\text{oz}}{1\,\text{cup}} =$

   b. $\dfrac{3\,\text{gal}}{1} \times \dfrac{8\,\text{qt}}{1\,\text{gal}} \times \dfrac{2\,\text{pt}}{1\,\text{qt}} \times \dfrac{2\,\text{cup}}{1\,\text{pt}} \times \dfrac{4\,\text{oz}}{1\,\text{cup}} =$

   c. $\dfrac{3\,\text{gal}}{1} \times \dfrac{2\,\text{qt}}{1\,\text{gal}} \times \dfrac{8\,\text{pt}}{1\,\text{qt}} \times \dfrac{2\,\text{cup}}{1\,\text{pt}} \times \dfrac{2\,\text{oz}}{1\,\text{cup}} =$

   d. $\dfrac{3\,\text{gal}}{1} \times \dfrac{4\,\text{qt}}{1\,\text{gal}} \times \dfrac{2\,\text{pt}}{1\,\text{qt}} \times \dfrac{8\,\text{oz}}{1\,\text{cup}} =$

   Answer: _____

## Lesson 14 Content Check

1. Using dimensional analysis, convert 45 tsp. to oz.

   Answer: _____

2. Using dimensional analysis, convert 459 g to lb.

   Answer: _____

3. Using dimensional analysis, determine how many 40 mcg tablets are needed to fill an ordered dose of 20 mcg.

   Answer: _____

4. Using dimensional analysis, determine how many milliliters it takes to get 5 mg of a 10 mg/5 mL solution.

   Answer: _____

5. Using dimensional analysis, determine how many milliliters it takes to get 25 mg of a 150 mg/3 mL solution.

   Answer: _____

6. Using dimensional analysis, determine how many milliliters a 2 mg dose of a 5 mg/mL solution would be.

   Answer: _____

7. Using dimensional analysis, determine how many 10 mg tablets would be needed to fill an ordered dose of 40 mg.

   Answer: _____

8. Using dimensional analysis, convert 100 mL to gtt (using 1 mL = 15 gtt)

   Answer: _____

9. Using dimensional analysis, convert 32 qt to mL.

   Answer: _____

10. Using dimensional analysis, convert 3 cups to gal.

    Answer: _____

11. Using dimensional analysis, convert 96 tbsp. to pt.

    Answer: _____

12. Using dimensional analysis, convert 47 mL to tbsp.

    Answer: _____

13. A patient needs 0.3 g of medication that is available in the pharmacy as a 750 mg/ 5 mL suspension. How many mL of the suspension would provide the patient with their appropriate dose? Use dimensional analysis to solve.

    Answer: _____

14. A prescriber ordered a dose of 60 mg of doxycycline. The medication is available as a 20 mg/2 mL solution. How many milliliters are needed to provide the ordered dose? Use dimensional analysis to solve.

    Answer: _____

15. A prescriber ordered a dose of 125 mg of Amoxil® suspension. The medication is available as a 500 mg/10 mL solution. How many milliliters are needed to provide the ordered dose? Use dimensional analysis to solve.

    Answer: _____

16. A prescriber ordered a dose of 80 mg of famotidine. The medication is available as a 40 mg/5 mL solution. How many milliliters are needed to provide the ordered dose? Use dimensional analysis to solve.

    Answer: _____

17. A prescriber ordered a dose of 100 mg minocycline oral solution. The medication is available in a 0.5 g/5 mL oral solution. How many milliliters are needed to provide the ordered dose? Use dimensional analysis to solve.

    Answer: _____

18. How many mL of a 3 g/10 mL solution is needed to get a dose of 1.5 g? Use dimensional analysis to solve.

    Answer: _____

19. How many grams of a drug with a strength of 5 g/500 mL will 0.2 L yield? Use dimensional analysis to solve.

Answer: _____

20. How many 0.5 mg tablets would be needed for a 250 mcg dose? Use dimensional analysis to solve.

Answer: _____

# LESSON 15

# Insulin and Other Unit Calculations

Please see the chart below to answer some follow-up questions:

| Type of Insulin | Examples | Strength & How Supplied | | | |
|---|---|---|---|---|---|
| | | 3 mL Pen | 3 mL Cartridge | 3 mL Vial | 10 mL Vial |
| Rapid-Acting | Lispro (Humalog®) | U-100 (KwikPen®) U-200 (KwikPen®) | U-100 | U-100 | U-100 |
| | Aspart (NovoLog®) | U-100 (FlexPen®) | U-100 | --- | U-100 |
| | Aspart (Fiasp®) | U-100 (FlexTouch®) | U-100 (Penfill®) | --- | U-100 |
| | Glulisine (Apidra®) | U-100 (SoloStar®) | U-100 | --- | U-100 |
| Short-Acting | Regular (Humulin® R, Novolin® R) | U-100 (KwikPen®, FlexPen®) U-500 (KwikPen®) | U-100 | U-100 | U-100 U-500 |
| Intermediate-Acting | NPH (Humulin® N, Novolin® N) | U-100 (KwikPen®, FlexPen®) | U-100 | U-100 | U-100 |
| Long-Acting | Glargine (Lantus®) | U-100 (SoloStar®) | --- | --- | U-100 |
| | Glargine (Basaglar®) | U-100 (KwikPen®) | --- | --- | --- |
| | Glargine (Toujeo®) | U-300 (SoloStar®) – 1.5 mL ONLY | --- | --- | --- |
| | Detemir (Levemir®) | U-100 (Flextouch®) | --- | --- | U-100 |
| | Degludec (Tresiba®) | U-100 and U-200 (FlexTouch®) | --- | --- | U-100 |

(*continued*)

UNIT 2: Introduction to Units and Problem-Solving Methodology

| Type of Insulin | Examples | Strength & How Supplied | | | |
|---|---|---|---|---|---|
| | | 3 mL Pen | 3 mL Cartridge | 3 mL Vial | 10 mL Vial |
| Pre-Mixed [Combines specific amounts of intermediate-acting and short-acting insulin in one bottle or insulin pen. The numbers following the brand name indicate the percentage of each type of insulin in order of %N/%R.] | Humulin® 70/30 | U-100 (KwikPen®) | U-100 | U-100 | U-100 |
| | Novolin® 70/30 | U-100 (FlexPen®) | --- | --- | U-100 |
| | Novolog® 70/30 | U-100 (FlexPen®) | --- | --- | U-100 |
| | Humalog® Mix 50/50 | U-100 (KwikPen®) | --- | --- | U-100 |
| | Humalog® Mix 75/25 | U-100 (KwikPen®) | --- | --- | --- |

*Highlighted insulins are available OTC!

1. What are the 5 types of insulin? _____
   _____
   _____

2. What are the ways in which insulin is supplied (and the quantity of each)?
   _____
   _____
   _____

## Insulin Calculations

Notice that for the strength of each insulin, there is the letter "U" followed by a dash and a number. This represents the amount of units per milliliter in that particular strength. Therefore:

1. U-100 has the relationship of 1 mL = _____ *The majority of insulin calculations
2. U-200 has the relationship of 1 mL = _____
3. U-300 has the relationship of 1 mL = _____
4. U-500 has the relationship of 1 mL = _____

**LESSON 15:** Insulin and Other Unit Calculations

5. How can we prevent medication errors related to accidentally picking the wrong strength, or using the wrong strength for calculations? _____
_____
_____

In order to solve problems involving insulin calculations, either the ratio-proportion method or the dimensional analysis method can be applied.

> **Example:** A patient requires 30 units of U-200 insulin. How many milliliters should the pharmacist instruct them to draw up in a syringe?

Using ratio-proportion:

$$\frac{1\,\text{mL}}{200\,\text{units}} = \frac{X\,\text{mL}}{30\,\text{units}} \quad \text{where } X = \boxed{0.15\,\text{mL}}$$

Using dimensional analysis:

$$\frac{30\,\text{units}}{1} \times \frac{1\,\text{mL}}{200\,\text{units}} = \boxed{0.15\,\text{mL}}$$

**STOP AND PRACTICE:** Convert the following units to milliliters:

1. 50 units of U-100

    Answer: _____

2. 70 units of U-300

    Answer: _____

3. 3 units of U-100

    Answer: _____

4. 22 units U-200

    Answer: _____

**STOP AND PRACTICE:** Convert the following milliliters to units:

1. 0.2 mL of U-300

    Answer: _____

2. 0.9 mL of U-100

    Answer: _____

3. 0.45 mL of U-100

    Answer: _____

4. 1.2 mL of U-100

    Answer: _____

## Milliequivalent and Unit Calculations

Many texts treat milliequivalent and unit calculations as separate entities – this is not necessary. Follow the guidelines and logic of either the ratio-proportion method or the dimensional analysis method to come to right answer.

> **Example**: How many milliliters of a 20 mEq/5 mL solution would be required for a 40 mEq dose?

Using ratio-proportion:

$$\frac{20\,\text{mEq}}{5\,\text{mL}} = \frac{40\,\text{mEq}}{X\,\text{mL}} \quad \text{where } X = \boxed{10\,\text{mL}}$$

Using dimensional analysis:

$$\frac{40\,\text{mEq}}{1} \times \frac{5\,\text{mL}}{20\,\text{mEq}} = \boxed{10\,\text{mL}}$$

**STOP AND PRACTICE:** Calculate the following:

1. A vial of potassium sulfate says that it's concentration per mL is 5 mEq. How many mL would it take to obtain a 50 mEq dose?

   Answer: _____

2. A prescriber orders 75,000 units of penicillin to be administered intramuscularly. The vial contains 50,000 units/mL penicillin. How many milliliters should be given?

   Answer: _____

3. A patient needs 20 mEq of potassium chloride. The pharmacy has 10 mEq tablets in stock. How many tablets should the patient receive for each dose?

   Answer: _____

4. Enoxaparin 80 units is prescribed for a patient to self-inject once daily for 7 days following a surgery. If the pharmacy only carries 100 units/mL auto-injector pens, how many mL will the patient need to inject themselves with to obtain the correct dosage?

   Answer: _____

5. If a vial of potassium chloride says that its concentration per mL is 0.45 mEq, how many mL would it take to get a dose of 6.75 mEq?

   Answer: _____

# LESSON 15: Insulin and Other Unit Calculations

## Sampling the Certification Exam:

1. Toujeo® insulin is available only as a 1.5 mL U-300 pen. How many units are in 1 mL of Toujeo®?

    a. 100 units

    b. 200 units

    c. 300 units

    d. 500 units

    Answer: _____

2. A patient is prescribed 40 mEq of K-Dur® BID. The pharmacy only stocks 20 mEq tablets on the shelf. How many tablets would the patient need to take in order to obtain the correct dose?

    a. ½ tablet

    b. ¼ tablet

    c. 1 tablet

    d. 2 tablets

    Answer: _____

3. You have a vial of heparin with a concentration of 1,000 units / mL. A patient is given 1.7 mL during an emergency procedure. How much heparin did they receive?

    a. 1,000 units

    b. 1,700 units

    c. 2,000 units

    d. 2,700 units

    Answer: _____

4. 25 units of U-500 Humulin-R® is equivalent to

    a. 0.083 mL

    b. 0.05 mL

    c. 0.125 mL

    d. 0.25 mL

    Answer: _____

5. Epinephrine (5,000 units/3 mL) is prescribed to a patient to treat their anaphylactic reaction to a bee sting. If they administer 0.5 mL, how many units did they receive?

   a. 222 units

   b. 833 units

   c. 988 units

   d. 1,333 units

   Answer: _____

## Lesson 15 Content Check

1. Insulin is usually measured in _____

2. What is the conversion between insulin and mL?

   Answer: _____

3. A patient draws up 0.5 mL of U-100 insulin in a syringe. How many units is this?

   Answer: _____

4. If a patient has been told to use 30 units of Humulin® U-100 insulin, how many mL should the patient draw up in an insulin syringe?

   Answer: _____

5. A prescriber writes for 20 units of U-200 insulin. How many milliliters would this be equivalent to?

   Answer: _____

6. A patient brings their insulin syringe to the pharmacy and asks you to mark on it where they should draw up 50 units of their U-300 insulin. What milliliter on the barrel are you going to circle?

   Answer: _____

7. A patient uses 7 units of U-100 insulin in the morning and 11 units of U-100 at lunch and dinner. How many milliliters does the patient use in total per day?

   Answer: _____

LESSON 15: Insulin and Other Unit Calculations

8. A pharmacist pulls a 3 mL vial of U-500 insulin from the shelf. How many total units are contained in the vial?

Answer: _____

9. Which contains less volume – 20 units of U-100 insulin or 20 units of U-200 insulin?

Answer: _____

10. How many units are in 0.75 mL of U-300 insulin?

Answer: _____

11. You have on hand a 20mEq/10 mL vial of potassium chloride. How many mEq are in 2 mL?

Answer: _____

12. If a vial of magnesium sulfate says that its concentration is 100 mEq/50 mL, what is its concentration per mL?

Answer: _____

13. If a label states that a vial of potassium chloride has a concentration of 2 mEq/1 mL, how many mEq are in 10 mL?

Answer: _____

14. A prescriber orders 500,000 units penicillin to be administered intramuscularly. If a single vial contains 1,000,000 units/mL penicillin, how many milliliters should be given?

Answer: _____

15. A prescriber orders 125,000 units penicillin intramuscularly. In stock is a vial containing 50,000 units/mL penicillin. How many milliliters should be given?

Answer: _____

16. An elderly man with a blood clot in his leg is prescribed heparin 6000 units subcutaneously every 12 hours. On hand is 10,000 units per mL vials of heparin. How many milliliters should be administered for a single dose?

Answer: _____

17. How many total units are in a 10 mL vial of penicillin if the strength is 1,000,000 units/mL?

Answer: _____

18. How many milliequivalents are in 3 ounces of potassium chloride solution with a strength of 20 mEq/15 mL?

Answer: _____

19. How many milliliters are required for a 20 mEq dose of sodium bicarbonate if the vial on hand says the strength is 50 mEq/50 mL?

Answer: _____

20. How many milliliters would it take to get a dose of 500,000 units of nystatin using a suspension of 100,000 units per milliliter?

Answer: _____

# LESSON 16

# Compare and Contrast Methodology

| Topic | Ratio/Proportion | Dimensional Analysis |
|---|---|---|
| What do you start with? | | |
| What is the symbol between the two fractions? | | |
| How should the units be aligned? | | |
| To solve the equation, you must: | | |

1. **T / F:** Either method used will achieve the same answer.

    a. Therefore, the method <u>YOU</u> use comes down to _____

## Math Literacy

*Please answer the following questions to better hone your ability to read math problems.*

1. A 24 mg dose of a medication is ordered. The solution strength available is 12.5 mg in 1.5 mL. How many mL should the pharmacy technician prepare?

    a. What is the **known** variable (represented as a fraction with units) in this problem? _____

        i. How did you determine this? _____

    b. What is the **unknown** variable in this problem? _____

        i. How did you determine this? _____

        ii. **Y / N:** Does this match any of the units that the known variable is given in?

            1. **Y / N:** Is any further action required to use this variable as is in your set up?

    c. What unit does your answer need to be in? _____

        i. How did you determine this? _____

        ii. **Y / N:** Does this match any of the units that the known variable is given in?

            1. **Y / N:** Is any further action required to use this variable as is in your set up?

d. Solve:

   Answer: _____

2. A physician requires 0.4 mg of a medication. The drug label of this medication reads 1000 mcg in 2 mL. How many mL should the pharmacy technician prepare?

   a. What is the **known** variable (represented as a fraction with units) in this problem?
   _____

   i. How did you determine this? _____

   b. What is the **unknown** variable in this problem? _____

   i. How did you determine this? _____

   ii. Y / N: Does this match any of the units that the known variable is given in?

   1. Therefore, you have to _____ before you can plug it in to solve!

   c. What unit does your answer need to be in? _____

   i. How did you determine this? _____

   ii. Y / N: Does this match any of the units that the known variable is given in?

   1. Y / N: Is any further action required to use this variable as is in your set up?

   d. Solve:

   Answer: _____

3. How many liters of a Lactulose® solution would be required to obtain a 1 kg dose?

   NDC 0603-1378-58
   LACTULOSE
   SOLUTION, USP
   10 g/15 mL

   a. The picture of the label above provides the **(known / unknown)** variable in this problem.

   i. Why/how? _____

   ii. This is also referred to as the drug's _____

   iii. So, the known variable (with units) can be represented as a fraction that looks like this: _____

   b. What is the **unknown** variable in this problem? _____

   i. How did you determine this? _____

   ii. Y / N: Does this match any of the units that the known variable is given in?

1. Therefore, you have to _____ before you can plug it in to solve!

   c. What unit does your answer need to be in? _____

      i. How did you determine this? _____

      ii. **Y / N:** Does this match any of the units that the known variable is given in?

         1. Therefore, you have to _____ before you report your answer!

   d. Solve:

   Answer: _____

4. If a physician orders 20 mg of this drug, how many milliliters would the pharmacy technician draw up?

   ```
   AIN00214              Rx Only      Each mL contains: Furosemide 10 mg, Water
   NDC 36000-284-25                   for Injection q.s., Sodium Chloride for
                                      isotonicity, Sodium Hydroxide and, if
                                      necessary, Hydrochloric Acid to adjust pH
        FUROSEMIDE                    between 8.0 and 9.3
        INJECTION, USP                WARNING: Discard Unused Portion. Use
                                      Only If Solution Is Clear And Colorless,
                                      Protect From Light
         100 mg/10 mL                 Store at 20° to 25°C (68° to 77°F); excursions
          (10 mg/mL)                  permitted to 15° to 30°C (59° to 86°F) [See
                                      USP Controlled Room Temperature]
                                      Directions for Use: See Package Insert

        FOR IV OR IM USE              M. L. No. : G/28/1156

                                      B. No.   :
      10 mL SINGLE DOSE VIAL

   Manufactured for:                  Exp. Dt  :
   Baxter Healthcare Corporation
   Deerfield, IL 60015 USA
                                                                   2018-10-27
            Baxter
   ```

   a. What is the drug's strength? _____

      i. Why is there more than one "strength"? How are they related?

      _____

      _____

      _____

# UNIT 2: Introduction to Units and Problem-Solving Methodology

1. What is the total quantity (in mL) of this vial? _____
2. Therefore, a drug's strength compares EITHER _____ to _____ OR _____ to _____.

   ii. So, the known variable (with units) can be represented as a fraction that looks like this: _____

   b. What is the **unknown** variable in this problem? _____

   i. How did you determine this? _____

   ii. Y / N: Does this match any of the units that the known variable is given in?

   1. Y / N: Is any further action required to use this variable as is in your set up?

   c. What unit does your answer need to be in? _____

   i. How did you determine this? _____

   ii. Y / N: Does this match any of the units that the known variable is given in?

   1. Y / N: Is any further action required to use this variable as is in your set up?

   d. Solve:

   Answer: _____

5. A pharmacy technician is required to prepare a 20,000 unit dosage from a solution that has a strength of 15,000 units/5 mL. How many milliliters should the pharmacy technician draw up?

   a. What is the **known** variable (represented as a fraction with units) in this problem? _____

   i. How did you determine this? _____

   b. What is the **unknown** variable in this problem? _____

   i. How did you determine this? _____

   ii. Y / N: Does this match any of the units that the known variable is given in?

   1. Y / N: Is any further action required to use this variable as is in your set up?

   c. What unit does your answer need to be in? _____

   i. How did you determine this? _____

   ii. Y / N: Does this match any of the units that the known variable is given in?

   1. Y / N: Is any further action required to use this variable as is in your set up?

   d. Solve:

   Answer: _____

6. A prescription drug has a strength of 500 mg per 5 mL. If a 0.75 g dose is required, how many mL must be given?

   a. What is the **known** variable (represented as a fraction with units) in this problem?
   _____

      i. How did you determine this? _____

   b. What is the **unknown** variable in this problem? _____

      i. How did you determine this? _____

      ii. Y / N: Does this match any of the units that the known variable is given in?

         1. Therefore, you have to _____ before you can plug it in to solve!

   c. What unit does your answer need to be in? _____

      i. How did you determine this? _____

      ii. Y / N: Does this match any of the units that the known variable is given in?

         1. Y / N: Is any further action required to use this variable as is in your set up?

   d. Solve:

   Answer: _____

## Lesson 16 Content Check

1. A patient needs to take 3 oz. of medication, but only has a syringe that dispenses a tbsp. at a time. How many tbsp. would he need to take to equal his dose?

   Answer: _____

2. You have an order for 10 mL of a 500 mg/tsp of medication. How many mg is the patient receiving with this dose?

   Answer: _____

3. You have an order for 40 mg of phenytoin 100 mg/mL solution. How many mcL need to be dispensed to the patient?

   Answer: _____

4. A vial of medication states that its concentration is 0.2 g in 3 mL. A patient needs a 500 mg dose. How many mL of the medication will be necessary?

   Answer: _____

5. A patient has been given 200 mcg of levothyroxine. If the patient used a 0.5 mg /5 mL solution, how many mL did they take to obtain that dosage?

Answer: _____

6. You have an order for 100 mcg of levothyroxine. You have 0.05 mg tablets on the shelf. How many tablets will you need to complete the order?

Answer: _____

7. You have an order for 5 mg of amlodipine. You only have the 10 mg tabs in stock on your shelf. How many tablets will you need to complete the order?

Answer: _____

8. You have an order for 2 mg of Xanax®. You only have the 0.5 mg tablets in stock on your shelf. How many tablets will you need to complete the order?

Answer: _____

9. You have a vial of heparin with a concentration of 5,000 units /10 mL. A patient is given 2 mL during an emergency procedure. How much heparin did they receive?

Answer: _____

10. Tegretol® suspension comes in 450 mL bottles. How many 10 mL doses are in each bottle?

Answer: _____

11. A 1 pt bottle (480 mL) contains 275 mg. How many micrograms are in each milliliter?

Answer: _____

12. If each tablet of a medication contains 0.15 mg, how many tablets are needed to equal 450 mcg?

Answer: _____

13. If a patient takes 1 & ¾ tsp of a 150 mg/4 mL suspension, how many milligrams will the patient take?

Answer: _____

14. A patient with a prescription for Levaquin® 250 mg tablets comes into the pharmacy. You check your inventory and notice that you only have a solution of 0.75 g/10 mL. How many mL will the patient need for one 250 mg dose of medication?

Answer: _____

15. You have an order for 800 mg of Bactrim® DS. You only have a 500 mg/5mL suspension on the shelf. How many mL will you need to complete the order?

Answer: _____

16. A patient has a prescription for Augmentin® 875 mg. The pharmacy only has a suspension of 250 mg/5 mL. How many mL will the patient need for one dose?

Answer: _____

17. A prescriber writes an Rx for Zosyn® 1.6 g for a patient. The pharmacy has a 500 mg/5 mL vial. How many mL will the patient need for their dose?

Answer: _____

18. You have an order for 400 mcg of medication for a patient who weighs 32 lbs. On the shelf you have a 0.5 mg/mL solution of this medication. How much medication will the patient receive for their dose?

Answer: _____

19. You have an order for 50 mg of hydroxyzine. You have 10 mg tablets on the shelf. How many tablets are needed to complete the order?

Answer: _____

20. A prescription is given for 0.5 g of medication. This drug comes in 250 mg, 500 mg, and 750 mg strengths. Which strength should be used?

Answer: _____

## Unit 2 Content Review

Solve the following using either ratio-proportion or dimensional analysis. Use the opposite method to check your work.

1. Ordered: ciprofloxacin 500 mg
   On hand: ciprofloxacin 250 mg tablets
   How many tablets should the patient receive?

   Answer: _____

2. Ordered: prednisone 20 mg
   On hand: prednisone 5 mg tablets
   How many tablets should the patient receive?

   Answer: _____

3. Ordered: Biaxin® 1 g
   On hand: Biaxin® 500 mg capsules
   How many capsules should the patient receive?

   Answer: _____

4. Ordered: cephalexin 325 mg
   On hand: cephalexin 125 mg / 5 mL
   How many milliliters should the patient receive?

   Answer: _____

5. Ordered: Prozac® 10 mg
   On hand: Prozac® 20 mg / 5 mL
   How many milliliters should the patient receive?

   Answer: _____

6. Ordered: heparin 10,000 units
   On hand: heparin 15,000 units / 1 mL
   How many milliliters should the patient receive?

   Answer: _____

7. Ordered: acetaminophen 60 mg
   On hand: acetaminophen 80 mg / 0.8 mL
   How many milliliters should the patient receive?

   Answer: _____

8. Ordered: metformin 1 g
   On hand: metformin 500 mg tablets
   How many tablets should the patient receive?

   Answer: _____

9. Ordered: erythromycin 0.3 g
   On hand: erythromycin 200 mg / 5 mL
   How many milliliters should the patient receive?

   Answer: _____

10. Ordered: Levoxyl® 0.15 mg
    On hand: Levoxyl® 300 mcg tablets
    How many tablets should the patient receive?

    Answer: _____

11. Ordered: Robaxin® 0.4 g
    On hand: Robaxin® 150 mg / mL
    How many milliliters should the patient receive?

    Answer: _____

12. Ordered: Keppra® 1g
    On hand: Keppra® 500 mg tablets
    How many tablets should the patient receive?

    Answer: _____

13. Ordered: phenobarbital 30 mg
    On hand: phenobarbital 20 mg / 5 mL
    How many milliliters should the patient receive?

    Answer: _____

14. Ordered: Xanax® 0.5 mg
    On hand: Xanax® 0.25 mg tablets
    How many tablets should the patient receive?

    Answer: _____

15. Ordered: Cogentin® 1.5 g
    On hand: Cogentin® 0.5 g tablets
    How many tablets should the patient receive?

    Answer: _____

16. Ordered: diphenhydramine 25 mg
    On hand: diphenhydramine 12.5 mg / 5 mL
    How many milliliters should the patient receive?

    Answer: _____

17. Ordered: diphenoxylate 2.5 mg
    On hand: diphenoxylate 5 mg / 5 mL
    How many milliliters should the patient receive?

    Answer: _____

18. Ordered: Lasix® 40 mg
    On hand: Lasix® 20 mg / 5 mL
    How many milliliters should the patient receive?

    Answer: _____

19. Ordered: Depakote® 0.5 g
    On hand: Depakote® 125 mg tablets
    How many tablets should the patient receive?

    Answer: _____

20. Ordered: cimetidine 400 mg
    On hand: cimetidine 200 mg tablets
    How many tablets should the patient receive?

    Answer: _____

21. Ordered: atropine 0.3 mg
    On hand: atropine 0.4 mg in 0.5 mL
    How many milliliters should the patient receive?

    Answer: _____

22. Ordered: phenobarbital liquid 0.5 g
    On hand: phenobarbital 250 mg / mL
    How many milliliters should the patient receive?

    Answer: _____

23. Ordered: Neurontin® 300 mg
    On hand: Neurontin® 100 mg capsules
    How many capsules should the patient receive?

    Answer: _____

24. Ordered: Risperdal® 250 mcg
    On hand: Risperdal® 0.5 mg tablets
    How many tablets should the patient receive?

    Answer: _____

25. Ordered: Lopid® 1.2 g

    On hand: Lopid® 600 mg tablets

    How many tablets should the patient receive?

    Answer: _____

**Multiple Choice** - *Identify the choice that best completes the statement or answers the question.*

26. An IV additive order for potassium chloride 20 mEq is sent to the pharmacy. Potassium chloride is available as 2 mEq / mL solution in the pharmacy. How many mL should you be added to the IV solution?

    a. 4 mL

    b. 10 mL

    c. 12 mL

    d. 15 mL

    Answer: _____

27. Morphine sulfate 20 mg IM is ordered by a prescriber. Morphine 15 mg / mL is available in the pharmacy. How many mL should be injected to achieve the correct dose?

    a. 0.3 mL

    b. 0.6 mL

    c. 0.8 mL

    d. 1.3 mL

    Answer: _____

28. A technician wants to prepare 0.2 g of an IM medication from a vial with a strength of 400 mg per mL. How many mL are needed for this dose?

    a. 0.5 mL

    b. 0.8 mL

    c. 1.2 mL

    d. 2.2 mL

    Answer: _____

29. A pediatrician ordered amoxicillin 1 g PO bid. Amoxicillin is available as 400 mg / 5 mL solution. How many mL should be given for one dose?

   a. 5 mL
   b. 7.5 mL
   c. 9.5 mL
   d. 12.5 mL

   Answer: _____

30. A prescription for famotidine 30 mg comes through the pharmacy. A solution of famotidine on the shelf has a strength of 15 mg per 5 mL. How many mL should be given for the dose of medication?

   a. 2 mL
   b. 5 mL
   c. 8 mL
   d. 10 mL

   Answer: _____

31. A prescriber ordered aminophylline 0.4 g for an IV additive. Aminophylline is available as a 25 mg / mL solution. How many mL should the technician add to the IV solution?

   a. 6 mL
   b. 12 mL
   c. 16 mL
   d. 18 mL

   Answer: _____

32. A pharmacy technician needs potassium chloride 16 mEq for IV additive. Potassium chloride is available as 2 mEq / mL solution. How many mL should be added to the IV bag?

   a. 4 mL
   b. 8 mL
   c. 10 mL
   d. 16 mL

   Answer: _____

33. A patient is to take 40 units of U-100 insulin each day. How many mL will the patient draw up for each dose?

   a. 0.2 mL

   b. 0.4 mL

   c. 2.2 mL

   d. 4.2 mL

   Answer: _____

34. A pediatrician orders dexamethasone 6,000 mcg for a patient. Dexamethasone is available as 4 mg / mL solution. How many mL should be injected in order to achieve the correct dose?

   a. 0.5 mL

   b. 1.2 mL

   c. 1.5 mL

   d. 1.8 mL

   Answer: _____

35. A prescriber has a standing order of naloxone 350 mcg IM for adult patients who come into the ER for an overdose. The nursing station stocks naloxone in a 0.4 mg / mL solution. How many mL should be administered to get the correct dose?

   a. 0.3 mL

   b. 0.6 mL

   c. 0.9 mL

   d. 1.2 mL

   Answer: _____

36. An IV additive order for calcium gluconate 0.93 mEq is sent to the pharmacy. Calcium gluconate is available as a 0.465 mEq / mL solution. How many mL should the technician add to the IV solution?

   a. 1 mL

   b. 2 mL

   c. 3 mL

   d. 5 mL

   Answer: _____

## LESSON 16: Compare and Contrast Methodology

37. An IV additive order for sodium chloride 60 mEq is sent to the pharmacy. Sodium chloride is available as 4 mEq / mL. How many mL should be added to the IV solution?

    a. 5 mL
    b. 10 mL
    c. 15 mL
    d. 18 mL

    Answer: _____

38. A prescriber orders interferon alfa-2b, recombinant, 5 million units IM for a patient in the ICU. Interferon alfa-2b is available as 5 million units / 0.5 mL solution. How many mL should be given for one dose?

    a. 0.2 mL
    b. 0.3 mL
    c. 0.5 mL
    d. 0.7 mL

    Answer: _____

39. A pharmacy technician is to prepare a 1.8 mg dosage from a solution with a concentration of 2 mg in 3 mL. How many mL should be used?

    a. 1.5 mL
    b. 2.2 mL
    c. 2.7 mL
    d. 3.2 mL

    Answer: _____

40. A medication with a strength of 0.75 mg per mL is available in the pharmacy. How many mL are needed to prepare a 2 mg dose?

    a. 1.2 mL
    b. 1.5 mL
    c. 2.3 mL
    d. 2.7 mL

    Answer: _____

## UNIT 2: Introduction to Units and Problem-Solving Methodology

41. An order for 0.5 mg of IM fentanyl citrate is sent to the pharmacy. How many milliliters of a 250 mcg per mL strength vial is needed for this medication order?

    a. 1 mL

    b. 2 mL

    c. 3 mL

    d. 4 mL

    Answer: _____

42. A pediatrician orders ampicillin 250 mg IM STAT. Ampicillin is available as 0.5 g / mL solution for injection. How many mL should be given for the correct dose?

    a. 0.2 mL

    b. 0.5 mL

    c. 0.7 mL

    d. 1.2 mL

    Answer: _____

43. Cyanocobalamin 750 mcg is ordered as a one-time dose. It is available as 1,000 mcg per 1 mL solution in a single-use vial. How many mL should be administered to achieve the dose?

    a. 0.5 mL

    b. 0.8 mL

    c. 1.2 mL

    d. 1.5 mL

    Answer: _____

44. Penicillin G 400,000 units IM is ordered for a patient. The medication is available as a 1,200,000 units per 2 mL solution. How many mL should be given to achieve the correct dose?

    a. 0.2 mL

    b. 0.7 mL

    c. 1.2 mL

    d. 1.5 mL

    Answer: _____

45. Ketorolac 20 mg IM is ordered as a one-time dose for a patient in the ER. Ketorolac is available as 15 mg / mL. How many mL should the nurse draw up for injection?

   a. 0.5 mL

   b. 0.8 mL

   c. 1.3 mL

   d. 1.7 mL

   Answer: _____

46. A physician orders terbutaline 800 mcg for SC injection for a patient in the medical-surgical unit. Terbutaline is available as a 1 mg / mL solution. How many mL should be drawn up for the correct dose?

   a. 0.2 mL

   b. 0.5 mL

   c. 0.8 mL

   d. 1.2 mL

   Answer: _____

47. A pediatrician orders gentamicin 0.1 g IV for a patient in the PICU. It is available in the pharmacy as a 40 mg / mL solution. How many mL contains the correct dose?

   a. 1 mL

   b. 2 mL

   c. 2.5 mL

   d. 2.8 mL

   Answer: _____

48. A dosage of 300 mg of clindamycin was ordered for a patient. The strength available in the pharmacy is 0.4 g in 1.5 mL. How many mL should be given for the correct dose?

   a. 0.5 mL

   b. 1.1 mL

   c. 1.8 mL

   d. 2.3 mL

   Answer: _____

49. The pharmacy has 1 mL vials of heparin, which contain 5,000 units each. How many mL do you need for a dose of 1,100 units?

    a. 0.11 mL

    b. 0.22 mL

    c. 0.65 mL

    d. 2.2 mL

    Answer: _____

50. Convert the following

    a. 5 tsp -> cups

    Answer: _____

    b. 10 oz -> gal

    Answer: _____

    c. 1 qt -> mL

    Answer: _____

# UNIT 3

# Retail Pharmacy Math

# LESSON 17

# Prescription Reading

A prescription's signa (often abbreviated "sig) gives instructions to the patient on how they should take their medication properly. Prescribers use a common short-hand abbreviation system in either lowercase or uppercase letters to quickly write down their instructions that are then translated into what patients can understand by a pharmacy technician.

## Common Abbreviations and Sig Reference Chart

| | Abbreviation | Meaning |
|---|---|---|
| **Route of Administration** | ad ‡ | right ear |
| | as ‡ | left ear |
| | au ‡ | each ear |
| | od ‡ | right eye |
| | os ‡ | left eye |
| | ou ‡ | each eye |
| | po | by mouth |
| | SL | sublingually |
| | top | topical |
| | AAA | apply to affected area |
| | pr | per rectum (rectally) |
| | pv | per vagina (vaginally) |
| | inh | inhalation, or inhale |
| | EN | in each nostril |
| | per neb | by nebulizer |
| | per g button | by/through gastric button |
| | per ngt | by/through naso-gastric tube |
| | SC ‡, SQ ‡, subc, subq ‡ | subcutaneous |
| | IM | intramuscular |
| | IV | intravenous |
| | IVPB | intravenous piggyback |

| | Abbreviation | Meaning |
|---|---|---|
| **Dosage Form** | tab | tablet |
| | cap | capsule |
| | SR, XR, XL | slow/extended release |
| | sol | solution |
| | susp | suspension |
| | syr | syrup |
| | liq | liquid |
| | supp | suppository |
| | crm | cream |
| | ung, oint | ointment |
| | lot | lotion |
| | pow | powder |

| | Abbreviation | Meaning |
|---|---|---|
| **Timing of Administration** | bid | twice a day |
| | tid | three times a day |
| | qid | four times a day |
| | am | morning |
| | pm | afternoon or evening |
| | q | every |
| | hs ‡ | at bedtime |
| | prn | as needed for |
| | ac | before food/meals |
| | pc | after food/meals |
| | stat | immediately, now |
| | q_h, q_° | every _ hour(s) |
| | qd * | every day |
| | qod * | every other day |

| | Abbreviation | Meaning |
|---|---|---|
| **Measurement** | i, ii | one, two, etc. |
| | U * | unit |
| | ss | one-half ( ½ or 0.5 ) |
| | gtt | drop |
| | mL | milliliter |
| | tsp, 3 | teaspoon |
| | tbsp | tablespoon |

| | Abbreviation | Meaning |
|---|---|---|
| **Measurement, continued** | fl. oz, ℥ | fluid ounce |
| | L | liter |
| | mcg, μg ‡ | microgram |
| | mg | milligram |
| | g | gram |
| | gr | grain |
| | mEq | milliequivalent |
| | aa | of each |
| | ad | up to |
| | as, qs | quantity sufficient |
| | lb, # | pound |
| | **Abbreviation** | **Meaning** |
| **Other** | UD ‡, AD | as directed |
| | NR | no refill |
| | D/C ‡ | discharge or discontinue |
| | DAW | dispense as written |
| | DNE | do not exceed |
| | N/V | nausea/vomiting |
| | c/c | cough and congestion |
| | cc ‡ | mL OR with meals (depending on use) |
| | w/ or c̄ | with |
| | w/o | without |
| | D₅W | dextrose 5% in water |
| | NS | normal saline |
| | Sig | label or directions |
| | SOB | shortness of breath |
| | BP | blood pressure |
| | tuss | cough |
| | TAT | until all taken |
| | x | for |

*These sig codes are on the Joint Commission's (TJC) "Do Not Use" list due to the likelihood of them being misunderstood or mistranslated, but they are still important to know because they are unfortunately still widely used.

‡These sig codes are on the Institute of Safe Medication Practices' (ISMP) "Error Prone Abbreviations, Symbols, and Dose Designation" list due to misinterpretations involving harmful or potentially harmful medication errors according to MERP (Medication Errors Reporting Program). Though research has found links between these abbreviations and errors in the field, aspiring pharmacy technicians should still be aware of the existence and interpretation if they run into them.

## Sig Transcription:

All prescription short-hand sigs will translate into patient speak in this format:

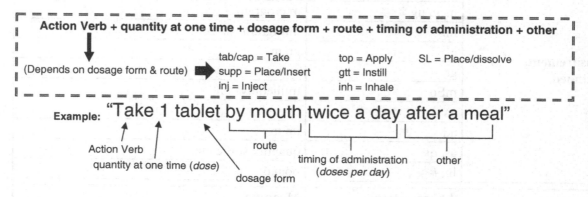

Example: "Take 1 tablet by mouth twice a day after a meal"

It is important to note that not everything a prescriber writes on a prescription hard copy has a short-hand abbreviation. In that case, a pharmacy technician must translate EVERYTHING a prescriber writes on the hardcopy as it is seen either in the order it is written or following the format above. It is OK to sometimes re-arrange things to make it make more sense, but the more direct translation, the more accurate the practice becomes. If there is ever a question as to the intended meaning of a sig, the prescriber must be contacted.

**STOP AND PRACTICE:** Translate the sigs below into patient speak.

1. 1 cap po qd - _____
2. 1 supp TID PR for pain - _____
3. 1 tab SL q8h prn nv - _____
4. 1.5 tab po q12h TAT - _____
5. 1 puff po q4-6h prn SOB/wheezing - _____

**STOP AND PRACTICE:** Convert the following patient labels into their corresponding sig codes:

1. Take 1 capsule by mouth every 3-4 hours as needed for pain.

2. Instill 2 sprays in each nostril every 12 hours as needed for allergic rhinitis.

3. Instill 2 drops into the left eye every 3-4 hours for 5 days.

4. Take 1 tablespoonful by mouth every 6 hours as needed for cough and congestion.

5. Apply 1 gram to the affected area topically twice daily.

## Best Practices:

Review the following sigs and interpretations that follow, then answer the follow-up questions to better discuss the best practices in sig translation for pharmacy technicians.

**Sig:** 1 tbsp po q12h

**Interpretation 1:** Take 1 tablespoonful by mouth every 12 hours.

**Interpretation 2:** Take 1 tablespoonful (15 mL) by mouth every 12 hours.

1. **(Y / N)** Is one "more correct" than the other?
2. What is the difference between the two? _____
3. What is the "best practice" when interpreting sigs for liquid drugs? _____

**Sig:** 1 supp PV qhs

**Interpretation 1:** Insert 1 suppository vaginally every night at bedtime.

**Interpretation 2:** Unwrap and insert 1 suppository vaginally every night at bedtime.

1. **(Y / N)** Is one "more correct" than the other?
2. What is the difference between the two? _____
3. What is the "best practice" when interpreting sigs for suppositories? _____

**Sig:** 50 units SQ QAM

   **Interpretation 1:** Inject 50 units subcutaneously every morning.

   **Interpretation 2:** Inject 50 units under the skin every morning.

   1. (Y / N) Is one "more correct" than the other?
   2. What is the difference between the two? _____
   _____
   _____

   3. What is the "best practice" when interpreting sigs for SQ/SC/subc/subq prescriptions? _____
   _____
   _____

**Sig:** 1 cap po ac and hs

   **Interpretation 1:** Take 1 capsule by mouth before meals and at bedtime.

   **Interpretation 2:** Take 1 capsule by mouth before a meal and at bedtime.

   1. (Y / N) Is one "more correct" than the other?
   2. What is the difference between the two? _____
   _____
   _____

   3. What is the "best practice" when interpreting sigs using the "ac" or "pc" shorthand? _____
   _____

       a. Why? _____
       _____
       _____

**Sig:** ½ tab po BID

   **Interpretation 1:** Take ½ tablet by mouth twice daily.

   **Interpretation 2:** Take 0.5 tablet by mouth twice daily.

   1. (Y / N) Is one "more correct" than the other?
   2. What is the difference between the two? _____
   _____
   _____

3. What is the "best practice" when interpreting sigs using partial numbers? ___
_____
_____

All pharmacy technicians must know how to accurately translate sig codes, but those who wish to follow best practices will try to make the sigs as easily understood by patients as possible. Sometimes, that means adding a little bit to a sig translation and/or rearranging the wording that a physician wrote out. Pharmacists, more senior technicians and personal experience over time will make this process easier but keeping the understanding of the patient in mind while translating is the best guide to effective prescription processing.

## Sampling the Certification Exam:

1. Match the following directions to the correct sig:

   Instill 1 drop into the left ear twice a day for 5 days
   a. 1 gtt AS BID x 5 days
   b. 1 gtt OS BID x 5 days
   c. 1 gtt AD BID x 5 days
   d. 1 gtt AU BID x 5 days

   Answer: _____

2. Match the following sig to the correct directions: 1 supp PV qod hs
   a. Insert 1 suppository rectally every other day at bedtime
   b. Insert 1 suppository vaginally every day at bedtime
   c. Insert 1 suppository vaginally every other day at bedtime
   d. Insert 1 suppository rectally every 8 days at bedtime

   Answer: _____

3. Which of the following is the sig code for the left eye?
   a. OU
   b. AU
   c. OS
   d. AS

   Answer: _____

4. Which of the following is the sig code for after meals?
   a. pc
   b. ac
   c. hs
   d. prn

   Answer: _____

5. Which of the following is the correct abbreviation for "until all are taken"?
   a. DAW
   b. NPO
   c. ss
   d. TAT

   Answer: _____

## Lesson 17 Content Check

1. Match the sig abbreviation to its correct route of administration:

   a. ad            _____ sublingually
   b. ou            _____ intramuscularly
   c. pv            _____ subcutaneously
   d. SL            _____ vaginally
   e. po            _____ both eyes
   f. subq          _____ topically
   g. IM            _____ right ear
   h. top           _____ intravenously
   i. IV            _____ by mouth

2. Match the sig abbreviation to its correct timing of administration:

   a. BID           _____ every 2 hours
   b. TID           _____ twice daily
   c. QID           _____ three times daily
   d. q             _____ every
   e. am            _____ as needed
   f. hs            _____ before meals

g. prn  \_\_\_\_ every day/once daily
h. ac   \_\_\_\_ four times daily
i. stat \_\_\_\_ morning
j. qod  \_\_\_\_ bedtime
k. q2h  \_\_\_\_ immediately
l. qd   \_\_\_\_ every other day

3. Match the sig abbreviation to its intended meaning:
   a. qs   \_\_\_\_ no refill
   b. UD   \_\_\_\_ quantity sufficient
   c. DAW  \_\_\_\_ as directed
   d. n/v  \_\_\_\_ nausea and vomiting
   e. NS   \_\_\_\_ normal saline
   f. NR   \_\_\_\_ dispense as written
   g. ung  \_\_\_\_ ointment

4. Match the following sigs to the questions below:
   a. 1 tab po QID
   b. 1 tab po TID
   c. 1-2 caps po q4h prn n/v
   d. 1-2 caps po q6h prn pain
   e. 1 supp pr q8h
   f. 1-2 gtt OU TID
   g. 1-2 gtt AS BID
   h. 1 tab SL BID prn headache
   i. 2 sprays in each nostril qd
   j. AAA prn itching

   \_\_\_\_ Which of these sigs more than likely refers to a topical cream?
   \_\_\_\_ Which of these sigs tells the patient to use their prescription for their eyes?
   \_\_\_\_ Which of these sigs tells the patient to use their prescription for their ears?
   \_\_\_\_ Which of these sigs tells the patient to use their prescription rectally?
   \_\_\_\_ Which of these sigs tells the patient to take their prescription daily?

____ Which of these sigs refers to a tablet to be taken four times a day?

____ Which of these sigs refers to a tablet to be taken three times a day?

____ Which of these sigs refers to a capsule for nausea and vomiting?

____ Which of these sigs refers to a capsule to be taken every 6 hours?

____ Which of these sigs tells the patient to take their prescription under the tongue?

5. Interpret the following into patient speak:

   a. i tab po qam

   b. ii tabs po today, then i tab po qd for 5 days

   c. i cap po TID pc

   d. i-ii tabs po q4h prn pain

   e. ii gtt OU q8h x 3 days

   f. inject 50 mcg SC qweek

   g. ii tabs po qd

   h. i tab po qd ac and hs

   i. i-ii inh po prn cough

   j. apply 1 patch qam. Remove hs.

6. Rewrite the following in pharmacy shorthand:
   a. Take 3 capsules by mouth at bedtime

   b. Inhale 1-2 puffs by mouth three times a day

   c. Instill 1 drop in the left ear as needed for pain

   d. Inject the contents of one syringe every other day

   e. Take 2 tablets by mouth every 4-6 hours as needed for nausea and vomiting

   f. Insert 1 suppository rectally three times a day as needed for pain

   g. Take 2 tablets by mouth now, then one tablet by mouth four times a day until all taken

   h. Place one tablet under the tongue every 4 hours for nausea

   i. Take one tablet by mouth before meals and at bedtime

   j. Instill 2 drops in each eye every morning for one week

# LESSON 18

# Day Supply Calculations

The day supply of a prescription is how long the prescription will last the patient if they take it as prescribed. It is important that it is correctly calculated to avoid issues with future fills or the patient's insurance company.

## Rules:

1. Assume all months are 30 days. This allows for pharmacy technicians to eliminate ambiguity or the question of how many days are in the month at that time. It also eliminates the need to account for when the patient started their medication – they have a "month" of medication whether they get it on the 1st or the 22nd of the month!

    *Exception*: if the prescription is written in terms of weeks, rather than days, we assume there are 4 weeks (i.e., 28 days) in a month.

2. If a partial number ends up being the answer, round down to the nearest one's place! Partial days cannot be entered into most computer software systems.

3. Always assume patients will use the <u>maximum amount</u> of the drug as allowed by the sig.

## Solving - Choose one of the following:

1. Ratio/Proportion Method **OR** Dimensional analysis

    a. Known #1: Dose at one time (based on the sig)

    b. Known #2: Doses per day (based on the sig)

    c. *(Possibly)* Known #3: Other conversions you have previously memorized/are given in the problem

    d. Unknown: Quantity

2. Formula method:

$$\text{Day supply} = \frac{\text{Quantity}}{(\text{dose at one time} \times \text{doses per day})}$$

3. Visual method: Draw it out! (See Appendix)

> **Example:** 2 cap po TID, quantity #90 caps

$$\frac{90\,caps}{1} \times \frac{1\,dose}{2\,caps} \times \frac{1\,day}{3\,doses} = \boxed{15\,day}$$

> All examples in this text from this point forward will mainly utilize the dimensional analysis method.

Remember to use the principles of whatever method is chosen and let the logic guide the placement of units, and therefore, numbers.

**STOP AND PRACTICE:** For the sigs below, practice identifying the elements of a day supply problem by answering each question for A and B as a fraction with units, then solve for C by using any of the methods described above.

1. 1 cap po qd

    a. What is the dose at one time?

    Answer: _____

    b. How many doses are they taking per day?

    Answer: _____

    c. How long would a quantity of #30 caps last?

    Answer: _____

2. 1-2 tab po QID prn muscle spasm

    a. What is the dose at one time?

    Answer: _____

    b. How many doses are they taking per day?

    Answer: _____

    c. How long would a quantity of #40 tabs last?

    Answer: _____

3. 3 units U-100 insulin BID

    a. What is the dose at one time?

    Answer: _____

    b. How many doses are they taking per day?

    Answer: _____

LESSON 18: Day Supply Calculations

c. How long would a quantity of 3 mL last?

Answer: _____

4. 30 units U-200 insulin QAM

   a. What is the dose at one time?

   Answer: _____

   b. How many doses are they taking per day?

   Answer: _____

   c. How long would a quantity of 10 mL last?

   Answer: _____

5. 1.5 tsp po TID

   a. What is the dose at one time?

   Answer: _____

   b. How many doses are they taking per day?

   Answer: _____

   c. How long would a quantity of 6 oz. last?

   Answer: _____

## Special Circumstances:

Certain sig's will require more thought to calculate the day supply due to how they are written. See examples of noteworthy sigs and an explanation of each below:

> **Example 1:** What is the day supply for a prescription with the following sig: 4 caps STAT, 2 po BID TAT, Qty: 40

STAT means immediately, so the patient is to take 4 capsules as soon as they get the prescription. Therefore, 4 capsules should be taken away from the quantity before the day supply is calculated.

40 – 4 = 36 capsules

$$\frac{36\,caps}{1} \times \frac{1\,dose}{2\,caps} \times \frac{1\,day}{2\,doses} = \boxed{9\,days}$$

> **Example 2:** What is the day supply for a prescription with the following sig: 1 tab po q4-6h, Qty: 60

Remember, when calculating sigs with variable timing or dose options, base all calculations off of the maximum use of the drug. In this case, every 4 hours is the most often.

$$\frac{60 \text{ tabs}}{1} \times \frac{1 \text{ dose}}{1 \text{ tab}} \times \frac{4 \text{ hours}}{1 \text{ dose}} \times \frac{1 \text{ day}}{24 \text{ hours}} = \boxed{10 \text{ days}}$$

> **Example 3:** What should the day supply be for a prescription with the following sig: 1-2 gtt OU BID, Qty: 5 mL (Assume for this drug, 1 mL = 15 gtt)

With all ear and eye drop medications, you MUST be given the conversion between mL and gtt to solve, and you have to pay close attention to the dose – is it for one eye/ear or both? In this example, 1 dose is 2 drops PER eye, so 1 dose is 4 drops in total.

$$\frac{5 \text{ mL}}{1} \times \frac{15 \text{ gtt}}{1 \text{ mL}} \times \frac{1 \text{ dose}}{4 \text{ gtt}} \times \frac{1 \text{ day}}{2 \text{ doses}} = 9.375 \text{ which rounds down to } \boxed{9 \text{ days}}$$

> **Example 4:** What should the day supply be for a prescription with the following sig: 2 caps po BID x 5 days, Qty: 20

Certain sig codes will be written very specifically in this way where the day supply is mentioned in the sig. You can always double-check the math to make sure it works out correctly (and that is encouraged), but the day supply can be easily seen by looking directly at the sig.

$\boxed{5 \text{ days}}$

Calculation to double check:

$$\frac{20 \text{ caps}}{1} \times \frac{1 \text{ dose}}{2 \text{ caps}} \times \frac{1 \text{ day}}{2 \text{ doses}} = \boxed{5 \text{ days}}$$

## Sampling the Certification Exam:

1. What is the day supply for the following prescription: 1 po BID, qty: 30
    a.  10 days
    b.  12 days
    c.  15 days
    d.  30 days

Answer: _____

**LESSON 18:** Day Supply Calculations

2. What is the day supply for the following prescription: 1 po q4-6h, qty: 120

   a. 10 days
   b. 20 days
   c. 30 days
   d. 40 days

   Answer: _____

3. What is the day supply for the following prescription: Inj. 10 units U-200 subQ QAM, qty: 3 mL

   a. 15 days
   b. 20 days
   c. 30 days
   d. 60 days

   Answer: _____

4. What is the day supply for the following prescription: 1 gtt AU BID, qty: 5 mL (use 1mL = 20 drops)

   a. 13 days
   b. 20 days
   c. 25 days
   d. 50 days

   Answer: _____

5. What is the day supply for the following prescription: ¼ tsp. po BID, qty: 100 mL

   a. 30 days
   b. 40 days
   c. 50 days
   d. 60 days

   Answer: _____

## Lesson 18 Content Check

1. What is the day supply for the following prescription: 1 po QAM & QPM, qty: 180

   Answer: _____

2. What is the day supply for the following prescription: 2 po TID, qty: 90

   Answer: _____

3. What is the day supply for the following prescription: 1 tabs po BID, qty: 90

   Answer: _____

4. What is the day supply for the following prescription: 2 gtt OU TID, qty: 10mL (use 1mL = 20 drops)

   Answer: _____

5. What is the day supply for the following prescription: 1 gtt AD BID, qty: 5 mL (use 1mL = 15 drops)

   Answer: _____

6. What is the day supply for the following prescription: 1 po q2°, qty: 20

   Answer: _____

7. What is the day supply for the following prescription: 4 tabs po STAT, then 2 po QID TAT, qty: 60

   Answer: _____

8. What is the day supply for the following prescription: Inj. 30 units U-100 SUBQ qam, qty: 10 mL vial

   Answer: _____

9. What is the day supply for the following prescription: 1-2 gtt OU BID, qty: 5 mL (use 1 mL = 20 gtt)

   Answer: _____

**LESSON 18:** Day Supply Calculations

10. What is the day supply for the following prescription: AAA 2-4 g BID prn pain, qty: 100 g

    Answer: _____

11. What is the day supply for the following prescription: 2 po QID, qty: 120

    Answer: _____

12. What is the day supply for the following prescription: 1 caps po BID × 2 days, 2 po TID × 3 days, 2 po QID × 5 days, then stop, qty: QS

    Answer: _____

13. What is the day supply for the following prescription: 1 tsp. po BID, qty: 8 oz.

    Answer: _____

14. What is the day supply for the following prescription: 1 puff TID, qty: 1 (1 inhaler = 200 puffs)

    Answer: _____

15. What is the day supply for the following prescription: 1 supp. PV qod hs, qty: 7

    Answer: _____

16. What is the day supply for the following prescription: 1 tsp po on day 1, then ½ tsp po on days 2-5, qty: 15 mL

    Answer: _____

17. What is the day supply for the following prescription: Inject 1.5 mL IM every 2 weeks, qty: 9 mL

    Answer: _____

18. What is the day supply for the following prescription: 1 po QPM, qty: 28

    Answer: _____

# UNIT 3: Retail Pharmacy Math

19. What is the day supply for the following prescription: Swish and spit 10 cc po TID prn burning mouth syndrome, qty: 300 mL

   Answer: _____

20. What is the day supply for the following prescription: 1 tab po BID on days 16-28 of cycle, qty: 36

   Answer: _____

# LESSON 19

# Quantity Calculations

Quantity sufficient (QS) is the number of doses of drug that will last the amount of time indicated by a prescription. In other words, the prescriber expects the pharmacy to calculate it.

## Rules:
1. The day supply must be known in order to calculate.
2. If a partial number ends up being the answer, round up to the nearest one's place! Pharmacies do not dispense partial tablets to patients, and other dose forms are impossible to split due to how they are manufactured.
3. Certain medications will always have the same quantity, and therefore, will be calculated and/or communicated differently.

> **Example:** Birth control packs almost always have 28 tablets in each pack

4. Always use the metric system when determining the quantity. Most computer software systems utilize the metric system for the data entry field where quantity is input.

## Solving - Choose one of the following:
1. Ratio/Proportion Method **OR** Dimensional analysis
   a. Known #1: Dose at one time (based on the sig)
   b. Known #2: Doses per day (based on the sig)
   c. *(Possibly)* Known #3: Other conversions you have previously memorized/are given in the problem
   d. Unknown: Day supply
2. Formula method:

> Quantity sufficient = Dose at one time × doses per day × days indicated

a. You may find that this method works VERY well for sigs that have varying dosages throughout the sig (ex: sigs for the drug prednisone, or warfarin).

3. Visual method: Draw it out! (See Appendix)

> **Example:** For the sig, 1-2 po QID prn muscle spasm, how many tablets would they need to fill the prescription for a 5-day supply?

$$\frac{5 \text{ days}}{1} \times \frac{4 \text{ doses}}{1 \text{ day}} \times \frac{2 \text{ tablets}}{1 \text{ dose}} = \boxed{40 \text{ tabs}}$$

Remember to use the principles of whatever method is chosen and let the logic guide the placement of units, and therefore, numbers.

**STOP AND PRACTICE:** For the sigs below, answer each question using any of the methods described above.

1. For the sig, ½ po qd, how many tablets would they need to fill the prescription for a two week's supply?

    Answer: _____

2. For the sig, 1 po qod, how many tablets would they need to fill the prescription for a 12 week's supply?

    Answer: _____

3. For the sig, 2 po TID, how many capsules would they need to fill the prescription for a 10-day supply?

    Answer: _____

4. For the sig, 3 units U-100 insulin BID, how many 3 mL pens would they need to fill the prescription for a 28-day supply?

    Answer: _____

5. For the sig, 1.5 tsp po TID, how many mL would be needed to fill the prescription for a week's supply?

    Answer: _____

## Special Circumstances:

Certain sigs will require more thought to calculate the quantity needed due to how they are written. See examples of noteworthy sigs and an explanation of each below:

> **Example:** How many tablets should be dispensed for a prescription with the following sig: 2 tabs po TID × 3 days, then 1 tab po BID × 3 days, then 1 tab po QD × 3 days, then stop. Qty: QS

This sig is an example of what a taper-down dose looks like. It is likely for the drug prednisone, or some other steroid that patients must slowly wean themselves down from. In this case, the formula method works the best. It is also helpful to break up the sig into individual sections of unique sig codes (usually, every time there is a comma), then calculate the quantity needed in each part independently, and finally add up the quantities to equal the total quantity needed.

| Sig Part | Formula | Quantity |
|---|---|---|
| 2 tabs po TID × 3 days, | (2 × 3 × 3) | 18 |
| then 1 tab po BID × 3 days, | (1 × 2 × 3) | 6 |
| then 1 tab po QD × 3 days, | (1 × 1 × 3) | 3 |
| + then stop | (0) | 0 |
| | | 27 tablets |

> **Example:** How many tablets should be dispensed for a prescription with the following sig: 1 tab po BID on days 1-14 of cycle; Qty: 1 month

This sig is referencing a menstrual cycle and in pharmacy practice, we always assume that a menstrual cycle lasts 28 days regardless of the person who requires the prescription. Therefore, the day supply would be 1 month, which is 28 days. The quantity calculations would require first determining the total number of days that the patient is actually taking the drug (in this example, 14 days), and then calculations can resume as normal.

$$\frac{14 \text{ days}}{1} \times \frac{2 \text{ doses}}{1 \text{ day}} \times \frac{1 \text{ tab}}{1 \text{ dose}} = \boxed{28 \text{ tablets}}$$

## Applications in the field:

Many times, pharmacy technicians will use quantity calculations to determine what drug to select for dispensing due to the variety of sizes of drug products. Some drug products cannot be split but you can give multiple units of the same quantity to equal the total that the patient needs, and there are other concerns to think of such as inventory control, insurance processing, and many other issues. Always use critical thinking when determining what is best and remember, you can always consult the institution's policy and procedure manual or ask a pharmacist/senior technician for help!

> **Example:** For a prescription with the following sig: 1 gtt AD TID, Qty: QS (Assume 1 mL = 20 gtt)
>
> If you have the choice between a 2.5 mL bottle, a 5 mL bottle, and a 7.5 mL bottle, which would be the best to choose to last the patient one month?

$$\frac{1 \text{ month}}{1} \times \frac{30 \text{ days}}{1 \text{ month}} \times \frac{3 \text{ doses}}{1 \text{ day}} \times \frac{1 \text{ gtt}}{1 \text{ dose}} \times \frac{1 \text{ mL}}{20 \text{ gtt}} = 4.5 \text{ mL needed}$$

Since we cannot give a partial bottle of eye/ear drops, the 5 mL bottle would be the best option.

Other times, physicians will write dosing amounts in a weight on the sig instead of the quantity and dose form. In this case, you must incorporate the strength of the drug into your calculations.

> **Example:** You have an order for azithromycin 100 mg po qd × 5 days. You have azithromycin 150 mg/5mL on the shelf. How many mL will the patient need to complete their prescription?

$$\frac{5 \text{ days}}{1} \times \frac{1 \text{ dose}}{1 \text{ day}} \times \frac{100 \text{ mg}}{1 \text{ dose}} \times \frac{5 \text{ mL}}{150 \text{ mg}} = 16.7 \text{ mL}$$

**STOP AND PRACTICE**: For the sigs below, answer each question using any of the methods described above.

1. A patient uses 12 units U-200 with breakfast, 15 units with lunch and 15 units with dinner. If a doctor wrote their prescription for a month supply and your inventory had both 3 mL vials and 10 mL vials in stock, which would be best and how much would you give to the patient?

   Answer: _____

2. A patient has a drug order that reads: 5 mg po qd on MWF, 7.5 mg po qd on TR, 2.5 mg po Sat and Sun. You have a 1 mg, 2.5 mg and 5 mg tablet on the shelf.

   a. Which tablet do you give to your patient and why?

   Answer: _____

   b. How many tablets would the patient with the above script need to last for a month supply?

   Answer: _____

3. You have an order for Amoxil® 250 mg po TID. You have Amoxil® 500 mg/5mL on the shelf. How many mL will the patient need for a ten-day supply?

   Answer: _____

4. A drug order reads: 5 mg po BID. You have a 10 mg tablet on the shelf.

   a. How many tablets is the patient taking at one time if you give them what you have available?

   Answer: _____

   b. How many tablets is the patient taking over the course of one day?

   Answer: _____

   c. How many tablets will the patient need to last them 30 days?

   Answer: _____

## Sampling the Certification Exam:

1. What quantity would be needed for the following prescription: 2 tabs po TID, qty: QS 1 month

   a. 120 tabs
   b. 150 tabs
   c. 180 tabs
   d. 240 tabs

   Answer: _____

2. What quantity would be needed for the following prescription: 1 cap BID, qty: QS 1 month

   a. 30 caps
   b. 60 caps
   c. 90 caps
   d. 120 caps

   Answer: _____

3. What quantity would be needed for the following prescription: 1.5 tabs po TID, qty: QS 1 month

   a. 90 tabs
   b. 110 tabs
   c. 125 tabs
   d. 135 tabs

   Answer: _____

4. What quantity would be needed for the following prescription: 2 caps po q4h, qty: QS 15 days

   a. 90 caps

   b. 120 caps

   c. 180 caps

   d. 210 caps

   Answer: _____

5. What quantity would be needed for the following prescription: 3 tabs po QID, qty: QS 3 months

   a. 360 tabs

   b. 540 tabs

   c. 810 tabs

   d. 1,080 tabs

   Answer: _____

## Lesson 19 Content Check

1. How many capsules are needed for the following prescription: T3C po TID, #QS 1 month.

   Answer: _____

2. How many tablets are needed for the following prescription: T2T po q6h, #QS 3 month

   Answer: _____

3. A prescriber orders Lantus® U-100 0.6 mL SQ daily for a patient. How many 3 mL pens would the patient's get to last him for 3 months?

   Answer: _____

4. You have an order for cefdinir 300 mg po q8h. You have cefdinir 500 mg/5mL on the shelf. How many mL should the patient receive for a 5-day supply?

   Answer: _____

# LESSON 19: Quantity Calculations

5. A patient takes 0.12 mL BID of U-500 Humalog®. How many units will he take over a 2-week period?

   Answer: _____

6. What quantity would be needed for the following prescription: 3 tabs po QID, qty: QS 3 months

   Answer: _____

7. What quantity would be needed for the following prescription: 1 tsp. po BID × 2 days, then ½ tsp. po qd × 3 days, qty: QS

   Answer: _____

8. What quantity would be needed for the following prescription: 1.5 tabs po on MWF, 2 tabs po TR, 1 tab po on Sat & Sun, qty: 1 month

   Answer: _____

9. What quantity would be needed for the following prescription: 1 oz. qd × 5 days, qty: QS

   Answer: _____

10. What quantity would be needed for the following prescription: 2 tabs po QID × 3 days, 2 tabs po BID × 3 days, 1 tab po QID × 3 days, 1 tab po BID × 2 days, 1 tab po qd × 2 days, ½ tab po qd × 4 days, qty: QS

    Answer: _____

11. What quantity would be needed for the following prescription: 1 tbsp. po BID, qty: QS 5 days

    Answer: _____

12. What quantity would be needed for the following prescription: 2 tabs po qd MWF, 1.5 tabs po qd TR, 2.5 tabs po qd Sat/Sun, qty: QS 1 month

    Answer: _____

13. What quantity would be needed for the following prescription: 5 tabs po qweekly, qty: QS 1 month

    Answer: _____

14. What quantity would be needed for the following prescription: 2 tsp po BID x 10 days, qty: QS

    Answer: _____

15. What quantity would be needed for the following prescription: 1-2 tbsp po TID ac, qty: QS 2 weeks

    Answer: _____

16. What quantity would be needed for the following prescription: 1-2 tab po q4h prn pain. DNE 8 tab/day, qty: QS × 3 weeks

    Answer: _____

17. What quantity would be needed for the following prescription: 1 tab po TID prn ADHD on school days, qty: QS 1 month

    Answer: _____

18. What quantity would be needed for the following prescription: Victoza 6 mg/mL (2 mL pen), sig: Inj 1.2 mg SQ qam, qty: 1 month

    Answer: _____

19. What quantity would be needed for the following prescription: Humulin U-100 Insulin, sig: 12 units TID with a meal, 2 units BID with a snack; 1-month supply Quantity in stock: 3 mL vial or 10 mL vial *(choose one)*

    Answer: _____

20. What quantity would be needed for the following prescription: Neomycin and polymixin ear drop (1 mL = 15 gtt), sig: 1 gtt AD QID × 10 days Quantity in stock: 5 mL, 7.5 mL and 10 mL vial *(choose one)*

    Answer: _____

# LESSON 20

# Refills and Total Pills

## Refills:

Technicians take refills over the phone from prescribers, nurses or medical assistants all the time as a service to our patients. This allows the patients to not have to go to the prescriber just for a refill of a maintenance medication. It is important to remember that a refill authorization is just the prescriber allowing the patient to get the EXACT same prescription they have been getting, and they will indicate if they want the patient to have any additional refills. Think of it as a "new prescription" plus possible additional refills

Be careful with the wording, and make sure to write the person's first name and last initial next to the refill authorization (if phoned in) or ensure that it was signed (if faxed in)!

1. "OK × #" versus "OK + #"

    a. If a refill is authorized "OK × #" a certain amount, the prescription will be renewed for 1 new fill with the #-1 for refills remaining

    > **Example:** "Ok × 3" is 1 new fill + 2 refills

    b  If a refill is authorized "OK + #" a certain amount, the prescription will be renewed for 1 new fill with the # indicated as is for refills remaining

    > **Example:** "Ok + 3" is 1 new fill + 3 refills

**STOP AND PRACTICE:** Indicate how many total fills the following refill authorizations will allow.

1. "OK + 3" = _____ new fill + _____ refills
2. "OK + 4" = _____ new fill + _____ refills
3. "OK × 5" = _____ new fill + _____ refills
4. "OK + 1" = _____ new fill + _____ refills
5. "OK" = _____ new fill + _____ refills

## Total Pills

This is the amount of drug that the patient will get over the entire span of their prescription once they have used all their refills. The quantity, the first fill and the number of refills remaining must be known in order to calculate.

> **Example:** How many total pills will the patient get over the course of their entire prescription for the following script: 1 po qd #30 tabs + 5 refills
>
> 30 tabs × 6 TOTAL fills (1 initial fill + 5 refills) = 180 tabs

**STOP AND PRACTICE:** Indicate how many total pills the patient will receive over the course of their entire prescription history for the following sigs:

1. 1 tab po qam, 1 tab po qpm #60 + 2 refills

    Answer: _____

2. 1-2 tab po q4h prn pain, #90 + 3 refills

    Answer: _____

3. 2 tab po TID, #QS 1 month + 4 refills

    Answer: _____

4. 1 tab po MWF, 2 po T/Th, 1.5 Sat/Sun, #QS 1 month + 3 refills

    Answer: _____

5. 3 tab po qweek, #QS 3 months + 1 refill

    Answer: _____

## Putting it All Together:

*Write out the following sig's in patient speak and answer the questions that follow.*

1. 1-2 tab po q4-6h prn pain: _____

    _____

    a. How many tablets is the patient taking in one dose?

    Answer: _____

**LESSON 20:** Refills and Total Pills

b. How many tablets is the patient taking in one day?

Answer: _____

c. How many tablets would the patient need for a one-month supply?

Answer: _____

d. If the patient was given a quantity of #40, how many days would this last?

Answer: _____

   i. If the patient was allowed to refill this prescription 5 times, how many total pills would the patient be taking over the course of their entire therapy?

Answer: _____

2. 1.5 tsp po BID × 2 weeks, qty: QS: _____
_____

a. How many mL is the patient taking in one dose?

Answer: _____

b. How many mL is the patient taking in one day?

Answer: _____

c. What is the day supply for this prescription?

Answer: _____

d. How many mL should the patient receive?

Answer: _____

e. How many refills would it take to get this patient 1 L of medication over the course of their entire therapy?

Answer: _____

3. 1 supp PV qhs on days 1-14 of cycle, qty: QS 1 month: _____
_____

a. How many suppositories is the patient taking in one dose?

Answer: _____

b. How many suppositories is the patient taking in one day?

Answer: _____

c. What is the day supply for this prescription?

Answer: _____

d. How many suppositories should the patient receive?

Answer: _____

e. If the patient gets 11 refills, how many total suppositories would the patient receive over the course of their entire therapy?

Answer: _____

4. 2 cap po TID × 2 days, then 1 cap po BID: _____
   _____

a. How many capsules is the patient taking in one dose?

Answer: _____

b. How many capsules is the patient taking in one day?

Answer: _____

c. How many capsules should the patient receive for a month supply?

Answer: _____

d. If the patient was given a quantity of #60, how many days would this prescription last?

Answer: _____

5. ½ tab po qd × 1 week, then 1 tab po qd × 1 week, then 1.5 tab po qd × 1 week, then 2 tabs po qd × 1 week. Maintain dosage thereafter. _____
_____
_____

a. How many tablets is the patient taking in one dose each week?

Answer: _____

b. How many total tablets is the patient taking each week?

Answer: _____

c. What does it mean to "Maintain dosage thereafter"? _____
_____

   i. How many tablets is the patient taking per day after they build up?

   Answer: _____

   ii. Why would a sig be written this way? _____
   _____

   iii. If this patient has 2 refills on this prescription, what should the sig say on the refills?
   _____

   1. How many tablets would they be getting each month?

   Answer: _____

   2. How many total tablets would be dispensed with those refills included?

   Answer: _____

## Sampling the Certification Exam:

1. A prescriber calls in a refill for a prescription stating "OK × 6." How many refills will the patient get from this authorization?

   a. 3
   b. 4
   c. 5
   d. 6

   Answer: _____

# UNIT 3: Retail Pharmacy Math

2. A patient brings in a prescription for 120 sucralfate tablets. If the physician circled 4 refills, how many total tablets would this patient get over the course of their entire prescription?

   a. 120 tablets

   b. 240 tablets

   c. 480 tablets

   d. 600 tablets

   Answer: _____

3. A patient calls the pharmacy asking how many L of medication she can get if she filled the entire quantity of her prescription because she is going out of the country and won't be able to refill it while she is gone. She states that her prescription says "Enulose® 10g/15 mL syrup #16 oz, 1 tbsp BID for 5 days to relieve constipation. Refills: 3". What should you tell her?

   a. 1.44 L

   b. 1.92 L

   c. 2.43 L

   d. 3.21 L

   Answer: _____

4. A prescriber calls in a refill for a prescription stating "OK + 4." How many total fills will the patient get from this authorization?

   a. 3

   b. 4

   c. 5

   d. 6

   Answer: _____

5. For a prescription with the following sig: 1 po QID, Qty: 120 Refills: 5

   How many total pills would the patient receive over the course of their entire therapy?

   a. 480

   b. 600

   c. 720

   d. 840

   Answer: _____

# Lesson 20 Content Check

1. A prescription is called in to the pharmacy in the following way by a prescriber in response to a refill request: "OK × 2." If the prescription in question originally allowed for #30 pills, how many total pills is the prescriber allowing?

   Answer: _____

2. A drug with #30 + 5 refills has a total of how many pills for the entire length of the prescription?

   Answer: _____

3. A prescription is called in to the pharmacy in the following way by a prescriber in response to a refill request: "OK × 3." If the prescription in question originally allowed for #60 pills, how many total pills is the prescriber allowing?

   Answer: _____

4. A drug with #20 + 2 refills has a total of how many pills for the entire length of the prescription?

   Answer: _____

5. A prescription is called in to the pharmacy in the following way by a prescriber in response to a refill request: "OK + 2." If the prescription in question originally allowed for #30 pills, how many total pills is the prescriber allowing?

   Answer: _____

6. A drug with #45 + 11 refills has a total of how many pills for the entire length of the prescription?

   Answer: _____

7. A prescription is called in to the pharmacy in the following way by a prescriber in response to a refill request: "OK + 3." If the prescription in question originally allowed for #60 pills, how many total pills is the prescriber allowing?

   Answer: _____

8. A drug with #10 + 1 refills has a total of how many pills for the entire length of the prescription?

   Answer: _____

9. A prescription is called in to the pharmacy in the following way by a prescriber in response to a refill request: "OK." If the prescription in question originally allowed for #45 pills, how many total pills is the prescriber allowing?

Answer: _____

10. A drug with #90 + 6 refills has a total of how many pills for the entire length of the prescription?

Answer: _____

11. For a prescription with the following sig: 5 tabs po qweek, Qty: QS 1 month, Refills: 5

    How many tablets will the patient receive over the course of their entire prescription?

Answer: _____

12. For a prescription with the following sig: 1 tab po q2h TAT, Qty: 2 weeks, Refills: 2

    How many tablets will the patient receive over the course of their entire prescription?

Answer: _____

13. For a prescription with the following sig: 1 tab po q4-6h prn pain, Qty: 3 weeks, Refills: 1

    How many tablets will the patient receive over the course of their entire prescription?

Answer: _____

14. For a prescription with the following sig: Inject 12 units U-100 qam with breakfast, 10 units with lunch, 15 units with dinner and 2 units for each snack (assume the patient eats 2 snacks per day), Qty: 1 month, Refills: 5

    How many 3 mL pens will the patient receive over the course of their entire prescription?

Answer: _____

15. For a prescription with the following sig: 30 units U-200 insulin QAM, Qty: 1 month, Refills: 3

    How many 10 mL vials will the patient receive over the course of their entire prescription?

Answer: _____

**LESSON 20:** Refills and Total Pills    **183**

16. For a prescription with the following sig: ½ tab po qd, Qty: 1 month, Refills: 11

    How many tablets will the patient receive over the course of their entire prescription?

    Answer: _____

17. A prescriber faxes in a refill authorization for "OK × 5." How many total fills will the patient get for their prescription?

    Answer: _____

18. A prescriber faxes in a refill authorization for "OK + 2." How many total fills will the patient get for their prescription?

    Answer: _____

19. A prescriber faxes in a refill authorization for "OK × 11." How many total fills will the patient get for their prescription?

    Answer: _____

20. A prescriber faxes in a refill authorization for "OK + 5." How many total fills will the patient get for their prescription?

    Answer: _____

# LESSON 21

# Measurements in the Lab

In order to discuss good lab practices and the math that is involved in determining accuracy of measurements, it is first important to discuss certain scientific properties of chemicals.

## Terminology:

Define the following –

Cohesion: _____

_____

Surface tension: _____

_____

Viscosity: _____

_____

Adhesion: _____

_____

Density: _____

_____

## Properties of Substances:

Perform the following experiment:

1. Draw up water into a syringe.

2. Drop the water onto the surface of a penny. Count the number of drops until the bubble formed over the penny bursts. Record your answer below.

3. Repeat the process with alcohol instead of water, making sure that the penny is fully dry before starting to count.

Drops of water: _____     Drops of alcohol: _____

185

1. Water molecules have very **high** cohesion and alcohol molecules have very **low** cohesion.

    a. Follow-up: Substances that have a very high cohesion means that the molecules within the substance

    **(stick together very well / don't stick together very well).**

2. Water has a very **high** surface tension, and alcohol has a very **low** surface tension.

    a. Follow-up: Substances with high cohesion will have a **(high / low)** surface tension, and conversely, substances with low cohesion will have a **(high / low)** surface tension.

3. Place the following chemicals in the appropriate box that best describes them: water, syrup, alcohol, honey

    | Low Viscosity | High Viscosity |
    |---|---|
    |  |  |

    a. Follow-up: In pharmacy practice, we refer to liquids by their type. If comparing a solution to a suspension, which would be more viscous and why? _____

    _____

    b. Follow-up: Substances with a very high viscosity will likely have a very **(high / low)** adhesion, and substances with a very low viscosity will likely have a very **(high / low)** adhesion.

4. Review the following graphics to see how cohesion and viscosity affect pharmacy practice:

    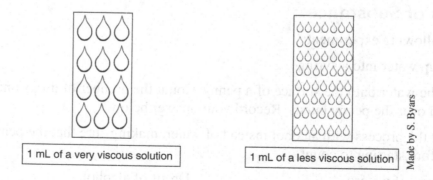

    a. Very viscous liquids have **(fewer / more)** drops in each mL whereas liquids with a low viscosity have **(fewer / more)** drops in each mL.

# LESSON 21: Measurements in the Lab

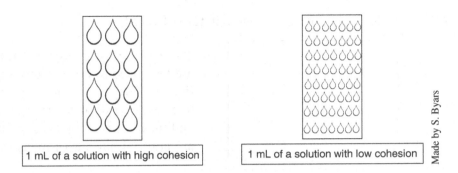

b. Liquids with high cohesion will have **(fewer / more)** drops in each mL whereas liquids with low cohesion will have **(fewer / more)** drops in each mL.

In pharmacy, density is measured as g/mL. Therefore, it is the only link between weight and volume that we have. We use the density of water (1 g/mL) as our standard.

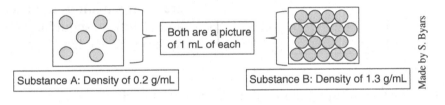

1. Why is the property of density significant in pharmacy practice? _____
   _____

2. Which substance above has the highest density? _____
   a. Why is that? _____

# Equipment:

Volumetric equipment/glassware in the lab will have one of the following important notations on them:

1. "TC", standing for "to contain".
   - This means the glassware is marked and contains a specified volume, but it <u>may not be accurate</u>; these pieces of equipment are best used to hold a chemical or mix chemicals with.

2. "TD", standing for "to deliver".
   - This means the glassware is calibrated to <u>accurately</u> deliver the stated volume; these pieces of equipment are the best for measuring out quantities needed

# UNIT 3: Retail Pharmacy Math

Examples of TC Glassware/Equipment – draw a picture of each!

1. Beaker:

   These pieces of glassware/equipment are best used to hold or mix solids or liquids. They are usually made with a lip to ease the task of pouring out a liquid, and they have calibration marks, but they are approximations of measurement.

2. Erlenmeyer Flask:

   These pieces of glassware/equipment are best used to mix liquids without spilling due to their high walls. Like beakers, they have calibration marks, but they are approximations of measurement and therefore, should not be used when accuracy is needed.

Examples of TD Glassware/Equipment – draw a picture of each!

1. Pipette:

   This is the most accurate piece of TD equipment for small volumes, but they are not widely used in mainstream pharmacy practice.

2. Syringe:

   This is a very accurate piece of TD equipment for small volumes that is also used widely in pharmacy practice. Additionally, there are various sizes available with multiple tip styles. The tips are either oral (aka: slip tip) or luer-lock, which is designed to safely attach a needle.

3. Volumetric Flask/ Cylinder:

This is the most accurate piece of TD equipment for large volumes, but they are not widely used in mainstream pharmacy practice.

4. Graduated Cylinder:

This is a very accurate piece of TD equipment for large volumes that is also used widely in pharmacy practice. Additionally, there are various sizes available and different materials available (glass and plastic).

## Measuring:

In order to accurately measure a liquid using TD glassware/equipment, it is important to understand the purpose and intent of calibration marks. These lines across the piece of equipment indicate the volume being measured when the liquid within reaches that mark. There is usually one at zero, one at the top (indicating the capacity of the equipment), and often, one directly in the middle, in addition to others evenly spaced out across the piece of equipment. Only the most significant measurements are marked with numbers, and the rest are left unmarked.

To understand how to read the measurements indicated at the unmarked calibration lines, follow these steps:

1. Choose 2 numbered calibration marks. Subtract the values – this indicates the volume between them.

2. Count the number of unmarked calibration lines between (and including) each one that was selected for step 1.

3. Divide step 1 by step 2 in order to determine the value of each unmarked line.

> **Example:** Determine the amount of the liquid in the following graduated cylinder:

Step 1) The calibration line above the liquid is 30 mL. The nearest marked calibration line below the liquid's surface is labeled 20 mL. Therefore, the gap is: 30 mL − 20 mL = 10 mL

Step 2) Counting the lines between and including the 20 mL and 30 mL calibration marks, there are a total of 10 lines.

Step 3) To determine the value of each line, take the total value and divide it by the number of lines between the marked lines.

Therefore: $\frac{10 \text{ mL}}{10 \text{ lines}} = 1$ mL/line

If each line represents 1 mL, and the liquid is sitting at the eighth line above the 20 mL mark, then the amount of liquid in the graduated cylinder is $\boxed{28 \text{ mL}}$.

**STOP AND PRACTICE:** Determine the value of each line in between the calibration marks on each of the following.

1.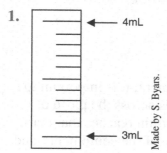

Answer: _____

2.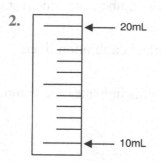

Answer: _____

3.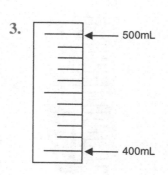

Answer: _____

No matter what, always use the smallest piece of equipment available to measure a liquid because the graduation marks will be the most accurate!

**STOP AND PRACTICE:** Determine to the best of your ability, where 1.3 mL can be found on the following pieces of equipment:

1. A 3 mL syringe:

   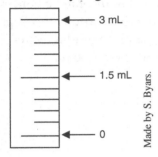

2. A 10 mL graduate cylinder:

   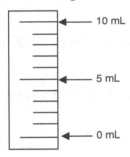

3. A 50 mL volumetric flask:

   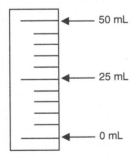

4. Which was easier? _____

   a. Why? _____
   _____

   b. Therefore: As the piece of volumetric equipment/glassware gets larger, the calibration markings become **(smaller in value / larger in value)**.

As a general rule, the volume being measured within a container should not be below 20% of the total capacity of the container.

In addition to understanding calibration marks, it is important to note how the properties of substances influence accurate measuring. Due to the various properties of a liquid (mainly cohesion, adhesion, surface tension and viscosity), a curve will form when measured out in a piece of glassware. This curve is called a meniscus. The meniscus of a liquid can be concave, or convex, and should be measured at the lowest/highest point in the curve (respectively) depending on which occurs. Additionally, since the curve can be misleading, the piece of glassware/equipment containing the liquid should be on a flat, level surface and should be read at eye level.

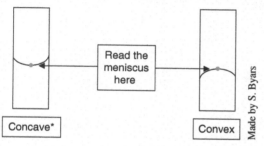

*Note that most chemicals you will encounter in pharmacy practice have concave meniscus'

1. Follow-up: Viscous liquids will have greater adhesion and therefore, a (**deeper** / **shallower**) meniscus than liquids with lower viscosity.

## Sampling the Certification Exam:

1. The most accurate piece of volumetric equipment commonly found in a pharmacy lab to measure a small volume is a:

   a. pipette

   b. volumetric cylinder

   c. syringe

   d. graduated cylinder

   Answer: _____

2. The most accurate piece of volumetric equipment commonly found in a pharmacy lab to measure a large volume is a:

   a. pipette

   b. volumetric cylinder

   c. syringe

   d. graduated cylinder

   Answer: _____

3. Which type of liquid is the most viscous?

   a. solution

   b. elixir

   c. suspension

   d. tincture

   Answer: _____

4. What is expressed with the units: g/mL?

   a. density

   b. molality

   c. molarity

   d. viscosity

   Answer: _____

5. A liquid in a concave piece of volumetric equipment is most accurately measured at the:

   a. top of the meniscus

   b. graduation mark

   c. middle of the liquid

   d. bottom of the meniscus

   Answer: _____

## Lesson 21 Content Check

1. Which type of glassware is more accurate: TD or TC?    Answer: _____
2. What is the main purpose of TC glassware? _____
3. What is the main purpose of TD glassware? _____
4. The two types of TD glassware used for measuring small volumes are: _____
   _____

   a. Which is the most common in pharmacy practice?    Answer: _____

5. The two types of TD glassware used for measuring large volumes are: _____
   _____

   a. Which is the most common in pharmacy practice?    Answer: _____

6. When measuring a liquid, you should always use the (**smallest / largest**) piece of equipment that is appropriate.

7. Draw an example of a liquid with a concave surface in a graduated cylinder:

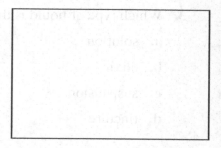

8. Draw an example of a liquid with a convex surface in an Erlenmeyer flask:

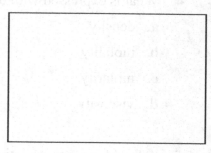

9. Determine the volume in the syringe:

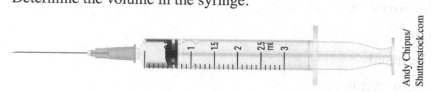

Answer: _____

10. Determine the volume in the syringe:

Answer: _____

11. Determine the volume in the syringe:

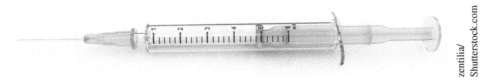

Answer: _____

12. Draw an arrow to where the following volumes could be found on this syringe:

   a. 2.3 mL
   b. 4.55 mL
   c. 8.1 mL
   d. 9.7 mL

13. Determine the volume in this graduated cylinder:

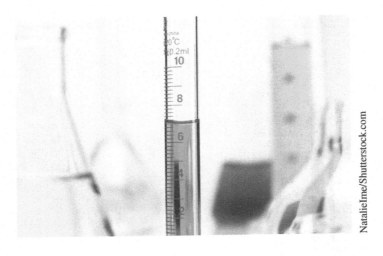

Answer: _____

UNIT 3: Retail Pharmacy Math

14. Determine the volume in this graduated cylinder:

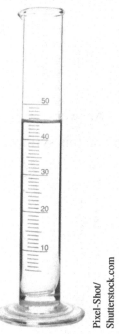

Answer: _____

15. Determine the volume in this beaker:

Answer: _____

# LESSON 22

# Accuracy and Percent Error

## Understanding Electronic Balances

The best electronic balances in pharmacy practice will measure out to the thousandths place, but some are less accurate. Regardless of the balance that is used, the last digit to the right is the least accurate digit of the measurement because the digit in the place value behind it cannot be determined.

> **Example:** An electronic balance weighing out a substance shows "1.362 g" in the measurement display window. The digit "2" in the thousandths place is the least accurate digit because the number in the ten thousandths place could be any digit from 0–9.

Electronic balances do not round measurements, but rather, cut off the number at the last place value where it can detect a change in measurement. This is known as the balance's sensitivity. Sensitivity of balances are measured in terms of "±", which means that the measurement can increase or decrease by a certain number and unit and the balance will detect that change only.

> **Example:** An electronic balance has a sensitivity of ± 5 mg. This means that measurements can only be accurate in increments of 5 mg because anything less than 5 mg cannot be detected by the balance.

**STOP AND PRACTICE:** Answer the following questions regarding sensitivity:

1. **(Y / N)** Can an electronic balance with a sensitivity of ± 10 mg measure out 57 mg accurately?

   a. Why/why not? _____

   _____

2. **(Y / N)** Can an electronic balance with a sensitivity of ± 3 mg measure out 3.3 g accurately?

   a. Why/why not? _____

   _____

3. (Y / N) Can an electronic balance with a sensitivity of ± 0.1 g measure a quantity of 3,420 mcg accurately?

   a. Why/why not? _____

   _____

4. An electronic balance has a sensitivity of ± 100 mg. What measurement would be reflected in the display window if you were attempting to measure out 1.434 g?

   Answer: _____

Accuracy is best defined as the degree to which the result of a measurement, calculation, or specification conforms to the correct value or standard. In pharmacy practice, a technician's goal should be 100% accuracy, but many things, such as human error, equipment maintenance and calibration, and distractions make that goal unfeasible. This is where the idea of error comes into the picture - error is the difference between what is *actually* measured, and what *should* have been measured. Percent error is a value that determines the scale of the error itself, giving the error meaning that can be understood by all. Saying that a measurement is off by 10 grams is meaningless because it does not give a clear indication of what should have been measured, but saying a measurement is off by 10% allows any audience to understand the amount of error involved. It is important to know error in order to calculate percent error and being able to perform these calculations helps pharmacy technicians determine how accurate their lab practices are.

## Calculating Error and Percent Error:

Formulas: $\text{Error (E)} = \dfrac{|\text{Quantity desired} - \text{Quantity actual}|}{\text{sensitivity of machine being used}}$  or  $\% E = \dfrac{E}{\text{Quantity desired}} \times 100\%$

Note: *The units on the numerator and the denominator must be the same!*

**Example:** What is the percent error if a pharmacy technician weighs out 5.2 g of a chemical instead of the 5.3 g that was needed?

Step 1) Find the error using the formula above.

$E = |5.3 \text{ g} - 5.2 \text{ g}| = 0.1 \text{ g}$

Note that error is always the absolute value (positive difference) between the quantity desired and the actual quantity weighed.

**LESSON 22:** Accuracy and Percent Error

Step 2) Plug in the error into the percent error formula and solve.

$$\% E = \frac{0.1 \text{ g}}{5.3 \text{ g}} \times 100\% = \frac{10}{5.3} = \boxed{1.89\%}$$

> Note that this set up allows us to cancel the unit of grams, leaving the % as the only unit left!

**STOP AND PRACTICE:** Answer the following questions regarding percent error:

1. What is the percent error if 100 mg of a substance is weighed on a balance with a sensitivity of ± 6 mg?

   Answer: _____

2. A pharmacy technician weighs 40 mg of menthol on a balance with a sensitivity of ± 5 mg. Calculate the percent error that may have occurred.

   Answer: _____

3. A 20 mL graduated cylinder is weighed and recorded as 85.49 grams. A technician uses the graduated cylinder to estimate 15 mL of water and finds that the combined weight of the graduated cylinder and the water is 100.28 grams. The technician knows that 15 mL of water should weigh only 15 grams. What is the technician's percent error?

   Answer: _____

4. A pharmacy technician is directed to weigh 15 g of substance and limit the percent error to no more than 1%. What is the maximum amount of error, in milligrams, that the technician would not be allowed to exceed?

   Answer: _____

## Least Weighable Quantity (LWQ):

The least weighable quantity on an electronic balance is the smallest amount that can be weighed within a certain degree of accuracy.

Formula: $\boxed{LWQ = \frac{E}{\%E} \times 100\%}$

Note: *The units of E and LWQ must be the same!*

# UNIT 3: Retail Pharmacy Math

> **Example:** What is the least weighable quantity in milligrams that can be weighed on an electronic balance with a sensitivity of ± 100 mg and a maximum potential error of 10%?

Step 1) Plug in the given values into the LWQ formula, ensuring that the units match, and solve!

$$\text{LWQ mg} = \frac{100 \text{ mg}}{10\%} \times 100\% = \frac{10{,}000}{10} = \boxed{1{,}000 \text{ mg}}$$

> Note that this set up allows us to cancel the unit of %, leaving mg as the only unit left!

**STOP AND PRACTICE:** Answer the following questions regarding least weighable quantity:

1. On a prescription balance with a sensitivity of ± 0.06 g, what is the least weighable quantity that can be weighed with a maximum potential error of not more than 12%?

   Answer: _____

2. On a prescription balance with a sensitivity of ± 5 mg, what is the smallest amount that can be weighed with a maximum percent error of 2.5%?

   Answer: _____

3. What is the least weighable quantity allowable for a prescription balance with a sensitivity of ± 0.01 g to allow for an error of no more than 5%?

   Answer: _____

## Applications:

Though patient instructions are often written in terms of household units (i.e., teaspoons or tablespoons), pharmacy technicians should ensure that every patient gets a medicine cup, dropper, or syringe with their prescription so no household spoons are used to measure medication. Common household utensils are widely inaccurate and should not be relied upon for exact dosing.

LESSON 22: Accuracy and Percent Error

## Sampling the Certification Exam

1. What is the percent error if 7.5 g of medication is weighed on a balance with a sensitivity of ± 6 mg?

    a. 0.08%

    b. 80%

    c. 8%

    d. 800%

    Answer: _____

2. What is the maximum error allowable on a weighable quantity of 25 grams with a 2% error?

    a. 2.5 g

    b. 1 g

    c. 50 g

    d. 0.5 g

    Answer: _____

3. On a prescription balance with a sensitivity of ± 3 mg, what is the smallest amount that can be weighed with a maximum percent error of 5%?

    a. 0.6 mg

    b. 75 mg

    c. 50 mg

    d. 60 mg

    Answer: _____

4. What is the potential percent error if 210 mg of medication is weighed on a balance with a sensitivity of ± 6 mg?

   a. 2.3%

   b. 2.9%

   c. 4%

   d. 3.4%

   Answer: _____

5. A technician weighed 75 mg of medication on a balance with a sensitivity of ± 2 mg. What is the percent error that might have occurred?

   a. 4%

   b. 2.7%

   c. 3%

   d. 1.5%

   Answer: _____

## Lesson 22 Content Check

1. What is the maximum error allowable on a weighable quantity of 25 grams with a 7% error?

   Answer: _____

2. What is the maximum error allowable on a weighable quantity of 75 grams with a 10% error?

   Answer: _____

3. What is the potential percent error if 511 mg of medication is weighed on a balance with a sensitivity of ± 5 mg?

   Answer: _____

4. What is the maximum error allowable on a weighable quantity of 100 grams with a 2% error?

   Answer: _____

LESSON 22: Accuracy and Percent Error    203

5. What is the potential percent error if 55 mg of medication is weighed on a balance with a sensitivity of ± 3 mg?

Answer: _____

6. What is the maximum error allowable on a weighable quantity of 15 grams with a 5% error?

Answer: _____

7. A technician weighed 95 mg of medication on a balance with a sensitivity of ± 4 mg. What is the percent error that might have occurred?

Answer: _____

8. Calculate the percent error for a measured weight of 185 mg with an ordered weight of 189 mg.

Answer: _____

9. What is the maximum error allowable on a weighable quantity of 102 grams with a 5% error?

Answer: _____

10. A technician weighed 300 mg of medication on a balance with a sensitivity of ± 3 mg. What is the percent error that might have occurred?

Answer: _____

11. On a prescription balance with a sensitivity of ± 2 mg, what is the smallest amount that can be weighed with a maximum percent error of 5%?

Answer: _____

12. A balance has a sensitivity of ± 20 mg. This means that the amount of change that the balance can detect can only be in increments of:

Answer: _____

13. A technician intends to measure 309 mg on a balance with a sensitivity of ± 20 mg. What is the percent error that might occur in this instance?

Answer: _____

14. What is the least weighable quantity on a balance that has a sensitivity of ± 10 mg and a known degree of error of 2.5%?

Answer: _____

15. Can an electronic balance with a sensitivity of ± 6 mg measure a quantity of 1.2 kg accurately?

Answer: _____

16. Can an electronic balance with a sensitivity of ± 12 mg measure a quantity of 2.4 lb accurately?

Answer: _____

17. Calculate the percent error for a measured weight of 13.381 g with an ordered weight of 13.4 g.

Answer: _____

18. What is the maximum error allowable on a weighable quantity of 60 milligrams with a 10% error?

Answer: _____

19. A technician intends to measure 41.5 mg on a balance with a sensitivity of ± 7 mg. What is the percent error that might occur in this instance?

Answer: _____

20. What is the least weighable quantity on a balance that has a sensitivity of ± 3 mg and a known degree of error of 4%?

Answer: _____

# LESSON 23

# Nonsterile Compounding

## Terminology:

Define the following –

USP: _____

_____

API: _____

_____

Master Log: _____

_____

Batch Log: _____

_____

## Using a Master Log:

Master logs contain information and techniques to help technicians through the process of making a specific compound. Each recipe is different and unique, so each compound will have a separate master log. They are set in their calculations for very specific quantities and rarely does a technician make the same amount that the Master log indicates. It is therefore the technician's job to determine the exact quantities of each ingredient needed.

Technicians can use the ratio/proportion method to determine these quantities:

> **Example:** For the following compound, how much of each ingredient will be required to make 300 mL?
> - Oseltamivir 75 mg capsule – 1
> - Cherry syrup – QS to 5 mL

This recipe has two ingredients; therefore, two separate calculations will be needed. Each ingredient's known ratio is the amount of the ingredient needed in the master log over the total quantity the master log makes. The unknown ratio will be the quantity of ingredient needed (the "*X*" in the proportion) over the quantity required for the prescription.

This compound's total quantity is 5 mL (as seen by the last ingredient where QS to 5 mL means to "add a sufficient quantity to reach 5 mL). Therefore, the calculations for each ingredient are as follows:

Oseltamivir:

$$\frac{1\,\text{cap}}{5\,\text{mL}} = \frac{X\,\text{cap}}{300\,\text{mL}} \quad \text{where } X = \boxed{60\,\text{caps}}$$

Cherry syrup:

No ratio is needed; instead, the answer will be $\boxed{\text{"QS to 300 mL"}}$

**STOP AND PRACTICE:** A technician is instructed to make 120 mL of a stomach suspension. When she pulls up the Master log, she sees the following information:

Acetaminophen – 2 g

Pantoprazole – 1.2 g

Water – to wet

Cherry syrup – QS to 50 mL

> Note that "to wet" quantities will never change; they are not measurable so no measurement will be specified.

1. What is the total quantity of suspension that this Master log will make?

   Answer: _____

2. What is the ratio of acetaminophen to total quantity in the master log?

   Answer: _____

3. What is the ratio of pantoprazole to total quantity in the master log?

   Answer: _____

4. How much of each ingredient are needed to make the full quantity of her suspension?

   Acetaminophen = _____

   Pantoprazole = _____

   Water = _____

   Cherry syrup = _____

## 1-2-3 Solution Calculations:

A 1-2-3 solution (sometimes also called "Magic Mouthwash") is a common mixture of:

- 1 part 2% Viscous Lidocaine
- 2 parts liquid Benadryl
- 3 parts Maalox (or Mylanta)

Most prescribers just give a total amount they want filled when writing/calling in for this prescription compound. It is up to the technician to determine how much of each ingredient to use, as is the case for most compounds. The solution is named 1-2-3 because it is 1 part 2% viscous lidocaine, 2 parts liquid Benadryl®, and 3 parts Maalox® (or Mylanta®). It is important to remember that liquids are additive, so 1 parts + 2 parts + 3 parts = 6 total parts. This means that each individual ingredient is:

$$\frac{1}{6} = 2\% \text{ Viscous Lidocaine}$$

$$\frac{2}{6} = \text{parts liquid Benadryl}^®$$

$$\frac{3}{6} = \text{parts Maalox}^® \text{ (or Mylanta}^®\text{)}$$

In order to determine the quantity of each ingredient to measure out for making the total solution in the right proportions, set each fraction up equal to $X$ / total amount of solution (in milliliters, or whatever unit is required).

2% Viscous Lidocaine

$$\frac{1}{6} = \frac{X \text{ mL}}{\text{Total mL}}$$  Where $X$ mL is the amount of 2% viscous lidocaine to measure out to add to

Liquid Benadryl®   +

$$\frac{2}{6} = \frac{X \text{ mL}}{\text{Total mL}}$$   $X$ mL of the liquid Benadryl®, to add to

Maalox® (or Mylanta®)   +

$$\frac{3}{6} = \frac{X \text{ mL}}{\text{Total mL}}$$   $X$ mL of the Maalox® (or Mylanta®).

**STOP AND PRACTICE:** Calculate the following 1-2-3 solution recipes:

1. If you needed to make 360 mL of a 1-2-3 solution, how many mL of each part would you need to add?

Answer: _____

2. If you needed to make 120 mL of a 1-2-3 solution, how many mL of each part would you need to add?

Answer: _____

3. If you needed to make 1 pint of a 1-2-3 solution, how many mL of each part would you need to add?

Answer: _____

## Acceptable Error:

USP <795> allows a ± 5% error on all compounded API quantities. As a quality assurance check, technicians often calculate their own percent error after a compound is made to ensure that it is within this threshold of acceptable error.

1. What should happen if the error is above or below the acceptable threshold? _____
_____
_____

Technicians should use the "quantity needed" as their target for 100% accuracy. Therefore, -5% would be 95% of the target goal, and +5% would be 105% of the target goal. Make sure to match the same unit as the ingredient quantities!

> **Example:** What is the acceptable range of error for a needed quantity of 150 mg of testosterone?

Minimum: $\dfrac{150\,mg}{100\%} = \dfrac{X\,mg}{95\%}$ where $X = 142.5$ mg

Maximum: $\dfrac{150\,mg}{100\%} = \dfrac{X\,mg}{105\%}$ where $X = 157.5$ mg

Therefore, the allowable range is $\boxed{142.5\,mg\text{–}157.5\,mg}$

**STOP AND PRACTICE:** Assume you made the following compounds and they were then sent to a lab for quality testing to determine how much API was present. Calculate the range of API allowed by USP that would be within the acceptable threshold when tested:

1. ibuprofen – 2.2 g

   Minimum Allowable Quantity: _____ Maximum Allowable Quantity: _____

2. ketamine – 300 mg

   Minimum Allowable Quantity: _____ Maximum Allowable Quantity: _____

# LESSON 23: Nonsterile Compounding

3. estradiol – 75 mcg

   Minimum Allowable Quantity: _____ Maximum Allowable Quantity: _____

4. testosterone – 6.75 g

   Minimum Allowable Quantity: _____ Maximum Allowable Quantity: _____

5. phenobarbital – 19,440 mg

   Minimum Allowable Quantity: _____ Maximum Allowable Quantity: _____

Because of this acceptable room for error, technicians usually calculate 10% over the required amount of API as indicated by the master log to ensure that the quantity of API will fall within the acceptable margin of error. Many compounds involve several steps in which the API will need to be transferred from one piece of equipment to another, and each time this occurs, some product is inevitably lost. The additional 10% helps to account for each loss of ingredient during transfer.

As before, technicians should use the "quantity needed" as their target for 100% accuracy. Therefore, 10% over the required amount would be 110% of the target goal. Make sure to match the same unit as the ingredient quantities!

> **Example:** How much API would a technician calculate to allow for the loss of transfer for a needed quantity of 150 mg of testosterone?

$$\frac{150\,mg}{100\%} = \frac{X\,mg}{110\%} \quad \text{where } X = \boxed{165\,mg}$$

**STOP AND PRACTICE:** Calculate how much API you would actually weigh/measure out if the master log said the quantity needed was the following:

1. 3.5 g

   Answer: _____

2. 2.78 L

   Answer: _____

3. 1.55 mcg

   Answer: _____

4. 21 mL

   Answer: _____

5. 120 mg

Answer: _____

## Sampling the Certification Exam:

1. How much more should be accounted for on the API of a nonsterile compound to make up for the amount lost during transfer?

    a. 3%

    b. 5%

    c. 8%

    d. 10%

    Answer: _____

2. A compound calls for 4.45 g of ibuprofen in the master log. If the USP wants to analyze your compound, what is the range of ibuprofen that can be measured and still be within the acceptable variance?

    a. 4.01 – 4.90 g

    b. 4.05 g – 4.75 g

    c. 4.23 g – 4.67 g

    d. 4.4 g – 4.5 g

    Answer: _____

3. You get a prescription for 8 ounces of a "1-2-3 Solution." How many mL of 2% Viscous Lidocaine will be needed?

    a. 40 mL

    b. 60 mL

    c. 80 mL

    d. 120 mL

    Answer: _____

4. How much variance does USP allow on the API of nonsterile compounds?

    a. ± 3%

    b. ± 5%

    c. ± 8%

    d. ± 10%

    Answer: _____

5. A compound calls for 3.872 g of acetaminophen in the master log. How much of the API will you measure out to account for loss during transfer?

   a. 3.99 g

   b. 4.066 g

   c. 4.151 g

   d. 4.259 g

   Answer: _____

## Lesson 23 Content Check

1. What is the range of variation upon quality testing allowed by USP on the following ingredients:

   a. rabeprazole 0.7 grams

   Answer: _____

   b. warfarin 5.9 grams

   Answer: _____

   c. levothyroxine 22.5 mg

   Answer: _____

   d. liothyronine 820 mcg

   Answer: _____

2. You are required to make the following prescription: Progesterone 400 mg Suppository, Qty: 24. When you pull the Master log, you see the following information:

   Master log - makes 1 suppository

        Progesterone Powder     0.4 g

        Polyethylene glycol 3350     1 g

        Polyethylene glycol 1000     3 g

   How much of each ingredient would be needed to make the prescription, accounting for the loss during transfer?

   Answer: _____

3. A prescription for 20 mL of Sun Ray Block solution comes into the pharmacy. When you pull the Master log, you see the following information:

   Master log – makes 100 mL of solution

   | Oxybenzone | 2 g |
   |---|---|
   | Hydrocortisone | 1 g |
   | 2% Viscous Lidocaine | 30 mL |
   | Aloe Vera Lotion | qs to 100 mL |

   How much of each ingredient would be needed to make the prescription, accounting for the loss during transfer?

   Answer: _____

4. A prescription comes into the pharmacy for the following: KBr 200 mg capsules, Qty: 50. When you pull the Master log, you see the following information:

   Master log – makes 300 capsules

   | Potassium bromide | 60 g |
   |---|---|
   | Methylcellulose | 32.5 g |
   | Lactose | 73.4 g |
   | #1 Red/Blue Caps | #300 |

   How much of each ingredient would be needed to make the prescription, accounting for the loss during transfer?

   Answer: _____

5. A prescription comes into the pharmacy for the following: Estriol 3 mg/mL cream, Qty: 12 mL. When you pull the Master log, you see the following information:

   Master log – makes 30 mL

   | Estriol | 0.09 g |
   |---|---|
   | Diethylene glycol | to wet |
   | Versabase | 29.8 mL |

   How much of each ingredient would be needed to make the prescription, accounting for the loss during transfer?

   Answer: _____

6. You are required to make #15 ounces of a 1-2-3 solution. How much of each ingredient, in mL, will be required?

Answer: _____

7. You are required to make 2 liters of a "Magic Mouthwash" solution. How much of each ingredient, in mL, will be required?

Answer: _____

## Unit 3 Content Review

**Multiple Choice** - *Identify the choice that best completes the statement or answers the question.*

1. A prescription reads "Synthroid® tablets 0.05 mg, 1 po b.i.d. #60." How many milligrams will the patient take over the course of one month?

    a. 3.0 mg

    b. 5.0 mg

    c. 30 mg

    d. 60 mg

    Answer: _____

2. A pharmacy technician received an order for sertraline 125 mg po qd. Sertraline is available in 50 mg scored tablets. How many tablets should be administered per day?

    a. 1 & ½ tabs

    b. 1 & 1/3 tabs

    c. 2 & ½ tabs

    d. 2 & 1/3 tabs

    Answer: _____

3. A prescription reads digoxin 125 mcg p.o. q.d. Digoxin is available as 0.25 mg / 5 mL elixir. How many mL of elixir are to be administered per day?

    a. 1.25 mL

    b. 1.50 mL

    c. 2.25 mL

    d. 2.50 mL

    Answer: _____

4. To decrease errors, what size graduate should be used to measure a specific quantity?
   a. less than 10 mL
   b. less than 100 mL
   c. the smallest available
   d. the largest available

   Answer: _____

5. To accurately read the amount of a substance inside a graduate, where should it be placed?
   a. in a refrigerator
   b. in a holding rack
   c. under a microscope
   d. on a flat, level surface

   Answer: _____

6. Which of the following is the definition of a meniscus?
   a. solute of a liquid
   b. solvent in a liquid
   c. the curve in the surface of a liquid
   d. layer between the diluent and solute

   Answer: _____

7. Insert information into the chart below to make it true for each row.

| Dose at one time: | Doses per day: | Quantity: | Day Supply |
|---|---|---|---|
|   | QID | 120 | 30 |
| 2 |   | 60 | 30 |
|   | Q6h | 60 | 15 |
| 1 |   | 90 | 30 |
|   | TID | 225 | 30 |
| 1.5 |   | 90 | 30 |

8. Write out the patient speak of the following sig abbreviations:
   a. gtt

   Answer: _____

   b. OD

   Answer: _____

c. AS

Answer: _____

d. pc

Answer: _____

e. qs

Answer: _____

f. hs

Answer: _____

g. qod

Answer: _____

h. QID

Answer: _____

**Interpret the prescribers' medication orders in patient speak and then answer the questions that follow:**

9. Hydroxyzine 25 mg, 1 tab po q6h prn agitation: _____
   _____

   a. How many tablets would the patient need for a 1-month supply?

   Answer: _____

   b. How long would #40 tablets last?

   Answer: _____

   c. If the pharmacy received a refill request back from the prescriber stating "OK + 6", how many total tablets would the patient be getting over the course of their entire prescription?

   Answer: _____

10. Hydrochlorothiazide 25 mg, 1 tab po bid: _____
    _____

    a. How many tablets would the patient need for a 3-month supply?

    Answer: _____

b. How long would #100 tablets last?

Answer: _____

c. If the pharmacy received a refill request back from the prescriber stating "OK + 1", how many total tablets would the patient be getting over the course of their entire prescription?

Answer: _____

11. Phenobarbital 60 mg, ss tab po tid: _____
_____

a. How many tablets would the patient need for a 5-day supply?

Answer: _____

b. How long would #50 tablets last?

Answer: _____

c. If the pharmacy received a refill request back from the prescriber stating "OK", how many total tablets would the patient be getting over the course of their entire prescription?

Answer: _____

12. Cheratussin® Elixir, 1 tsp po q4-6h prn cough: _____
_____

a. How many mL would the patient need for a week's supply?

Answer: _____

b. How long would #5 ounces last?

Answer: _____

c. If the pharmacy received a refill request back from the prescriber stating "OK x 5", how many total milliliters would the patient be getting over the course of their entire prescription?

Answer: _____

**LESSON 23:** Nonsterile Compounding

13. Prednisone 7.5 mg, 1 tab po qd x 5 days, then ss tab po qd x 3 days, then stop: _____
    _____

    a. What is the day supply for this sig?

    Answer: _____

    b. If the prescriber wrote for a quantity of "QS," how many tablets would be dispensed to the patient?

    Answer: _____

14. Penicillin G, 400,000 U IM q6h x 2 days: _____
    _____

    a. What is the day supply for this sig?

    Answer: _____

    b. If the prescriber wrote for a quantity of "QS," how many mL would be dispensed to the patient if this drug came in a vial of 200,000 U/10 mL?

    Answer: _____

15. Ancef®, 1 g IV q12h x 5 days: _____
    _____

    a. What is the day supply for this sig?

    Answer: _____

    b. If the prescriber wrote for a quantity of "QS," how many mL would be dispensed to the patient if this drug came in a vial of 500 mg/5 mL?

    Answer: _____

16. Cortisporin® otic suspension, 2 gtt AU tid x 5 days: _____
    _____

    a. What is the day supply for this sig?

    Answer: _____

b. If the prescriber wrote for a quantity of "QS," how many mL would be dispensed to the patient if this drug came in packages of 2.5 mL, 5 mL, and 10 mL (Assuming 1 mL = 20 gtt)?

Answer: _____

17. Assume patients eat 3 meals a day and 2 snacks a day:
   a. Inject 15 units U-100 qam and 10 units qpm: _____
   _____

   i. How many mL is the patient using in one day?

   Answer: _____

   ii. How many 3 mL vials would the patient need to fill a month supply?

   Answer: _____

   iii. If this insulin also came in a pen form, how many 3 mL pens would this patient need to fill a month supply?

   Answer: _____

   b. Inject 2 units U-100 qam, and 5 units TID with food: _____
   _____

   i. How many mL is the patient using in one day?

   Answer: _____

   ii. How many 3 mL vials would the patient need to fill a month supply?

   Answer: _____

   iii. If this insulin also came in a pen form, how many 3 mL pens would this patient need to fill a month supply?

   Answer: _____

   c. Inject 3 units U-100 with each meal, and 1 unit with each snack: _____
   _____

   i. How many mL is the patient using in one day?

   Answer: _____

ii. How many 10 mL vials would the patient need to fill a month supply?

Answer: _____

iii. If this insulin also came in a pen form, how many 3 mL pens would this patient need to fill a month supply?

Answer: _____

18. If the maximum potential error is + 60 mg in a total of 200 mg, what is the percentage of error?

Answer: _____

19. A prescriber orders 500 mg of a powdered medication. The amount is weighed, and the pharmacy technician double-checks it by using a more sensitive balance to weight it again. This procedure shows the amount to be 475 mg. The difference is 25 mg. What is the percentage of error?

Answer: _____

20. On a Class A prescription balance that is sensitive to 8 mg, what is the smallest quantity that can be weighted with a potential error of no more than 5 percent?

Answer: _____

21. A graduated cylinder was used by a technician to measure 20 mL of a medication. To double-check, the technician used a narrow-gauge burette, and found that 22 mL was actually measured. What is the percentage of error?

Answer: _____

22. A pharmacist prepared an ointment using 28.5 g zinc oxide instead of the 30.5 g that were required. Calculate the percentage of error on the basis of the desired quantity.

Answer: _____

23. What is the smallest amount of a substance you can weigh with an error not greater than 5% if the prescription balance has a sensitivity requirement of 0.004 grams?

Answer: _____

UNIT 3: Retail Pharmacy Math

24. What is the range of variation allowed by USP on the following ingredients?
    a. acetaminophen 6.6 grams

    Answer: _____

    b. nitroglycerin 0.3 grams

    Answer: _____

    c. digoxin 2.5 mg

    Answer: _____

    d. phentolamine 200 mcg

    Answer: _____

25. A "magic mouthwash" compound follows the same master log as the _____
    a. The ingredients are: _____
    _____

    b. If you wanted to make a quantity of #10 ounces, how many mL of each would be needed?

    Answer: _____

26. How much more or less, in milligrams, is it safe to weigh and fall within the 10% allowable range of error for a compound with an ingredient with the quantity needed of 2.4 grams?

    Answer: _____

27. Once a prescription suspension has been compounded, the technician has used a total of 6.5 g of API for the 200 mL bottle of medication.
    a. What is the strength of the medication in milligrams per mL?

    Answer: _____

    b. What is the strength of the medication in milligrams per teaspoonful?

    Answer: _____

# UNIT 4

# Pharmacy Business Math

# LESSON 24

# Inventory Management

Inventory is a measure of the products available for sale or use by the pharmacy. Proper maintenance of inventory is critical to running an efficient pharmacy. There will always be a simultaneous mixture of drugs ready for patient pick up, drugs being counted by pharmacy staff, drugs on the shelf for future use, and drugs that need to be ordered. To manage all these numbers, most computer software systems update the **b**alance **o**n **h**and (**BOH**) of every drug in real-time as prescriptions are processed through the pharmacy. Every drug has a **p**eriodic **a**utomatic **r**eplenishment (**PAR**) level (sometimes also referred to as the reorder point), or a minimum amount of drug that should be kept on-hand in the inventory at all time, and sometimes they will also have a maximum threshold that should not be exceeded. PAR levels are based on the use of a drug – the higher the use, the higher the PAR level. To help maintain the inventory, computer systems will automatically order drugs when their BOH falls below the PAR level. Pharmacy technicians must be sure that a drug's BOH is always correct so that the process of managing their inventory is simple, quick, and accurate.

When ordering a drug, it is important to know the package size of a product in addition to the PAR and BOH levels. All computer systems are different in how they measure package sizes, so every pharmacy technician should become familiar with their system before attempting to complete or check an order for accuracy. No matter how the information is presented, every drug will have a unit of measure (ex: tab, mL, pack) in order to give context and meaning to the quantity displayed. Additionally, <u>packages cannot be split,</u> so only whole numbers can be used when ordering a medication.

> **Example:** A pharmacy technician sees that the BOH of trazodone 100 mg is 62 tablets and the reorder point for this drug is 300 tablets. If this drug comes in a package of 100 tablets each, how many bottles should they order to fulfill their inventory needs?

Step 1) Determine if the BOH falls below the PAR level. If it does, the drug needs to be ordered. If it does not, then the drug does not need to be ordered.

   In this case, 62 is less than 300. An order will need to be placed.

Step 2) Determine how many bottles to order by comparing the package size to the BOH. Add these numbers together until the total reaches/exceeds the PAR level.

$$62 + 100 = 162 + 100 = 262 + 100 = 362 \text{ tablets}$$
$$\quad\quad (1) \quad\quad\quad (2) \quad\quad\quad (3)$$

> This pharmacy technician would need to order **3 bottles** of 100 tablets to meet their inventory needs.

**STOP AND PRACTICE:** Determine how many packages the computer system would order based on the following information:

| # | Drug | Package Size | PAR | BOH | Order |
|---|---|---|---|---|---|
| 1 | amoxicillin 500 mg cap | 500 | 240 | 180 | |
| 2 | anastrazole 1 mg tab | 30 | 90 | 62 | |
| 3 | Armour Thyroid® 1&1/2 gr tab | 90 | 120 | 180 | |
| 4 | atenolol 50 mg tab | 1,000 | 240 | 107 | |
| 5 | azithromycin 250 mg dosepack | 6 tabs x 3 packs | 18 | 7 | |
| 6 | benzonatate 200 mg cap | 100 | 180 | 120 | |
| 7 | buprenorphine 8 mg tab | 30 | 60 | 60 | |
| 8 | carbamazepine 400 mg ER cap | 100 | 400 | 315 | |
| 9 | carvedilol 12.5 mg tab | 500 | 600 | 398 | |
| 10 | clobetasol 0.05% cream | 15 g | 1 | -1 | |
| 10 | clobetasol 0.05% cream | 60 g | 1 | 1 | |
| 11 | cyanocobalamin 1 mg/mL vial | 1 mL x 20 vials | 15 | 4 | |
| 12 | desvenlafaxine ER 100 mg tab | 30 | 90 | 105 | |
| 13 | Depakote® 250 mg tab | 100 | 90 | 60 | |
| 14 | epinephrine 0.3 mg pen | 2 pens | 4 | 6 | |
| 15 | finasteride 1 mg tab | 30 | 90 | 72 | |
| 16 | furosemide 40 mg tab | 1,000 | 240 | 134 | |
| 17 | gabapentin 300 mg cap | 1,000 | 250 | 98 | |
| 18 | Humalog® U-100 KwikPen® | 3 mL x 5 pens | 15 | 9 | |
| 19 | Invokanna® 300 mg tab | 30 | 30 | 60 | |
| 20 | Janumet® XR 50 mg/1,000 mg tab | 30 | 60 | 120 | |
| 21 | Kariva® dosepack | 28 x 6 packs | 28 | 56 | |
| 22 | lactulose 10 g/15 mL | 473 mL | 1,419 | 950 | |

*(continued)*

## LESSON 24: Inventory Management

| #  | Drug                         | Package Size | PAR    | BOH    | Order |
|----|------------------------------|--------------|--------|--------|-------|
| 23 | latanoprost 0.005%           | 2.5 mL       | 7.5    | 2.5    |       |
| 24 | lisinopril 10 mg tab         | 1,000        | 240    | 146    |       |
| 25 | lisinopril 40 mg tab         | 100          | 90     | 152    |       |
| 26 | minocycline 50 mg cap        | 50           | 30     | 12     |       |
| 27 | methylprednisolone 4 mg dosepack | 21       | 63     | 0      |       |
| 28 | mupirocin 2% ung             | 22 g         | 2      | -2     |       |
| 29 | NuvaRing®                    | 3            | 3      | -2     |       |
| 30 | nystatin oral suspension     | 1 pt/bottle  | 480 mL | 720 mL |       |
| 31 | omega-3-acid 1 g             | 120          | 120    | 180    |       |
| 32 | ondansetron 4 mg ODT         | 30           | 60     | 45     |       |
| 33 | permethrin 5% cream          | 60 g         | 1      | 0      |       |
| 34 | ProAir® Inh.                 | 8.5 g        | 2      | 3      |       |
| 35 | rosuvastatin 40 mg tab       | 90           | 360    | 243    |       |
| 36 | scopolamine 1 mg patch       | 4            | 4      | 2      |       |
| 37 | sertraline 100 mg tab        | 500          | 600    | 198    |       |
| 38 | triamcinolone 0.025% cream   | 15 g         | 2      | 1      |       |
|    |                              | 80 g         | 2      | 0      |       |
| 39 | triamcinolone 0.1% ung       | 80 g         | 3      | 1      |       |
|    |                              | 1 lb         | 1      | 1      |       |
| 40 | zolpidem 10 mg tab           | 500          | 280    | 310    |       |

Consider this – when would it be acceptable for a pharmacy technician to greatly exceed the maximum on hand level when ordering? _____

_____

_____

**STOP AND PRACTICE:** Try the following word problems.

1. The PAR level for furosemide 20 mg is 200 tablets. The BOH says that 50 tablets are available in stock. Available sizes of medications with the cost for each size are as follows:

    30 tablets for $3.50

    90 tablets for $8.50

    100 tablets for $10.00

    500 tablets for $44.00

a. How many tablets should be ordered? _____

b. Determine the cost per tablet for each of the container sizes.

#30 = _____   #100 = _____

#90 = _____   #500 = _____

c. What would be the most cost-effective bottle? _____

2. The PAR level for digoxin 0.25 mg is 150 tablets and the BOH is 32 tablets. Available sizes of medications with the cost for each size are as follows:

   25 tablets for $6.50

   50 tablets for $11.75

   100 tablets for $19.25

   a. How many tablets should be ordered? _____

   b. Determine the cost per tablet for each of the container sizes.

   #25 = _____   #50 = _____   #100 = _____

   c. What would be the most cost-effective bottle? _____

      i. How many bottles should be ordered? _____

3. The PAR level for Eliquis® is 180 tablets. The pharmacy technician double checks the inventory on Monday because they know the computer will generate an order on Tuesday. The BOH for Eliquis® is 105 on Monday, and is available in containers of 90 tablets each.

   a. How many bottles of Eliquis® should be ordered? _____

   b. After the order is sent off, the same technician receives two prescriptions for Eliquis® - one for #45, and one for #90. How should the technician proceed with the filling of these prescriptions? Explain. _____
   _____
   _____

# Turnover

A turnover rate is a measure of how frequently a medication (or any product) is sold over time. Knowing the turnover rate can be useful for maintaining inventory because they help to establish the PAR levels.

Formula: | Turnover rate = Total purchases for a given time / Value of inventory at that time |

Notice that a specific time set must be indicated, and that there are no "units" to a turnover.

> **Example:** What is the annual turnover for a pharmacy who has an average inventory of $250,000 and spent $1,890,000 in purchasing that inventory?

Step 1) Plug in the numbers from the problem into the formula and solve:

Turnover rate = $1,890,000 / $250,000 = ☐ 7.56 ☐

This means that the entire inventory of that pharmacy (theoretically) turned over, or was sold, 7.56 times.

Follow-up:

1. An annual turnover of 3 means that it takes _____ months for the inventory to be completely sold and replenished.

2. An annual turnover of 16 means that it takes _____ weeks for the inventory to be completely sold and replenished.

3. A monthly turnover of 2 means that it takes _____ days for the inventory to be completely sold and replenished.

4. T / F : Turnover rate is only calculated on the inventory level as a whole, not on specific drugs.

    a. Explain: _____
    _____
    _____

5. What does a low turnover rate mean for the pharmacy? _____

    a. How can this possibly be adjusted/fixed? _____
    _____

6. What does a high turnover rate mean for the pharmacy? _____

    a. How can this possibly be adjusted/fixed? _____
    _____

**STOP AND PRACTICE:** Try the following word problems.

1. A pharmacy does a quarterly inventory and has an average of $100,000.00 on the shelf. The pharmacy's annual inventory purchases are $500,000.00. What is the pharmacy's turnover rate?

Answer: _____

2. If a pharmacy's average inventory for the past year was $132,936.00 and the annual cost total was $1,612,000.00, what was the turnover rate?

Answer: _____

3. If a pharmacy's average inventory for the past year was $156,200.00 and the annual cost total was $1,768,000.00, what was the turnover rate?

Answer: _____

## Sampling the Certification Exam:

1. Fluoxetine 20 mg capsules can be ordered in a bottle of 90 capsules. The maximum inventory level is 240 capsules and the minimum level is 60 capsules. If the current inventory is 25 capsules, how many bottles need to be ordered to reach the minimum?

   a. 3 bottles
   b. 2 bottles
   c. 1 bottle
   d. 0 bottles

Answer: _____

2. Minocycline 50 mg capsules can be ordered in bottles of 50 capsules. The maximum inventory level is 200 capsules and the minimum level is 30 capsules. If the current inventory is 27 capsules, how many bottles need to be ordered to reach the minimum?

   a. 4 bottles
   b. 3 bottles
   c. 2 bottles
   d. 1 bottle

Answer: _____

3. If a pharmacy's average inventory is $987.50 and the annual purchases are $7,538.00, what is the turnover rate?

   a. 13
   b. 12.7
   c. 7.8
   d. 7.6

Answer: _____

LESSON 24: Inventory Management     229

4. A _____ is a specific time period over which the total inventory is sold.

    a. sales period

    b. credit period

    c. turnover rate

    d. stocking rate

    Answer: _____

5. Turnover rate equals _____ divided by average inventory.

    a. total inventory

    b. total purchases

    c. wholesale prices

    d. total stock

    Answer: _____

## Lesson 24 Content Check

1. Order enough bottles to reach but not exceed the maximum:

| Drug and strength | Package Size | Min | Max | BOH | # to Order |
|---|---|---|---|---|---|
| amoxicillin 875 mg | 100 tabs | 20 | 120 | 12 | |
| Benicar® 40 mg | 30 tabs | 60 | 180 | 52 | |
| ciprofloxacin 500 mg | 100 tabs | 300 | 700 | 168 | |
| estradiol 0.01% cream | 42.5 g | 2 | 5 | 1 | |
| fluconazole 150 mg | 1 x 6 tab | 18 | 36 | 5 | |
| minocycline 100 mg | 60 cap | 20 | 70 | 18 | |
| Zyprexa® 10 mg | 30 tab | 10 | 30 | 11 | |

2. Calculate the turnover rate for the following drugs:

| Drug | Average Inventory | Annual Purchases | Turnover Rate |
|---|---|---|---|
| alprazolam 2 mg | $500 | $18,500 | |
| baclofen 10 mg | $370 | $12,980 | |
| carisoprodol 350 mg | $245 | $5,290 | |
| duloxetine 60 mg | $720 | $21,300 | |
| enalapril 5 mg | $180 | $3,003 | |

3. A drug has a minimum par level of 120 and a maximum par level of 500. Currently there are 72 pills in stock. If the drug comes in bottles of #50, how many bottles should be ordered to reach the minimum?

Answer: _____

4. What is the turnover for the drug ropinirole 2 mg with an average inventory of $1,290 and annual purchases of $15,700?

Answer: _____

5. What is the turnover for the drug Xanax® 2 mg with an average inventory of $2,155 and annual purchases of $11,350?

Answer: _____

6. The inventory of propranolol 120 mg is to be kept at a minimum of 300 and maximum of 1500. If 212 tabs are left at the end of the day, how many 100 ct bottles should be ordered to reach the minimum?

Answer: _____

7. If a pharmacy's average inventory for the past year was $234,000.00 and the annual cost total was $1,492,000.00, what was the turnover rate?

Answer: _____

8. If a pharmacy's average inventory for the past year was $89,556.00 and the annual cost total was $2.8 million, what was the turnover rate?

Answer: _____

9. What is the turnover for the drug amoxicillin 250 mg with an average inventory of $860 and annual purchases of $44,720?

Answer: _____

10. A drug has a minimum par level of 200 and a maximum par level of 600. Currently there are 43 pills in stock. If the drug comes in bottles of #90, how many bottles should be ordered to reach the minimum?

Answer: _____

# LESSON 25

# Income, Overhead, and Profit

## Terminology:

Fill in the following chart to better understand the definitions of the following:

| Term | Definition | Real Life Example | Synonym |
|---|---|---|---|
| Income | | | |
| Overhead | | | |
| Profit | | | |

## Discussion:

Any business that wants to stay functioning and relevant in the industry must maintain a certain profit line. Though pharmacies are ultimately concerned with quality healthcare and good patient outcomes, they are also a business and must make a profit to keep their operations functioning and improving. A basic equation can help anyone calculate an amount of profit.

Formula: $$\text{Profit (\$)} = \text{Income (\$)} - \text{Overhead (\$)}$$

A clear distinction must be understood between profit and percent profit. As stated above in the formula, profit is just the difference between the income and overhead for a business. Percent profit is a measure of how much was made over the expenses, and helps determine the scale of the profit itself, giving it meaning that can be understood by all. Saying that a business has a $1,000 profit is meaningless because it does not give the audience a full picture of the scale involved but saying that a business has a 2.5% profit allows anyone to understand more of that picture and scale. You must know the dollar amount of the profit to determine the percent profit.

> Follow up: When calculating percent profit, always compare it to the (**income / overhead**) which represents 100%.
>
>> Think about it: Any amount of money you bring in from your paycheck (income) that exceeds the cost of your bills (overhead) is money you can spend on whatever you want (profit).

Therefore, using the above logic:

1. **T / F:** A 5% profit can be theoretically represented by an income of 105%.

> **Example:** The independent pharmacy, Max's Emporium, would like to calculate how much percent profit was made after $595,000 in expenses were paid from their total income of $792,500.

Step 1) Calculate the $ of profit by using the formula above.

$792,500 − $595,000 = $197,500

   *Income*    *Overhead*    *Profit*

Step 2) Using the ratio-proportion method, calculate how much % profit this is over the overhead (which is 100%)

$$\frac{\$595{,}000}{100\%} = \frac{\$197{,}500}{x\%} \text{ where } x = \boxed{33.19\%}$$

> Remember to NOT use your calculator's % button as in this formula, % is a unit used to make sure things align as they should for the problem to be make logical sense.

**STOP AND PRACTICE:** Use the following expenses to determine your answers to the next few questions:

| | |
|---|---|
| Pharmacist salary | $120,000 |
| Technician salary | $90,000 |
| Rent | $45,000 |
| Utilities | $5,500 |
| Computer Maintenance | $2,500 |
| Software subscriptions | $1,250 |
| Liability Insurance | $5,000 |
| Business Insurance | $8,000 |
| Drug purchases | $900,000 |

1. What is the total overhead?

                                                                            Answer: _____

2. How much money is an 18% profit?

                                                                            Answer: _____

3. To make this goal, how much income must the pharmacy make?

Answer: _____

4. If the pharmacy only made $1,280,090.00 what is the percent profit?

Answer: _____

## Gross Profit and Net Profit

Gross profit, also known as markup or margin, represents the difference between what something is sold for and what it costs. This "profit" does not consider the expenses it took to host the product on the shelf or pay the staff to ring it up – that is where net profit comes in. In the pharmacy, all expenses are added up and divided by the average number of prescriptions processed through the system in a given time, giving a small fee that can be added to each prescription to help cover the cost of expenses. That fee is called either a dispensing fee or a professional fee.

Formulas:

> Gross Profit (GP) = sales price − acquisition cost

> Net Profit (NP) = GP − dispensing fee

> Dispensing Fee = total pharmacy overhead / total pharmacy prescriptions

**Example:** A drug is bought from the wholesaler for $23.55 for 100 tablets. The pharmacy decides to sell 30 tablets for $9.50. If a $0.75 dispensing fee is added onto the prescription at the time of sale, what was the pharmacies gross and net profit for that drug?

Step 1) Before any of the formula's above can be used, the cost of the 30 tablets that were sold must be determined. A simple proportion can accomplish this.

$$\frac{\$23.55}{100 \text{ tabs}} = \frac{\$x}{30 \text{ tabs}} \text{ where } x = \$7.07$$

Step 2) Plug the information into the formulas to determine each answer.

GP = $9.50 − $7.07 = $2.43

NP = $2.43 − $0.75 = $1.68

# UNIT 4: Pharmacy Business Math

**STOP AND PRACTICE:** From the following information, calculate the gross and net profit.

1. metformin 850 mg

    Pharmacy's Purchase Price: $37.99 per 100 tablets

    Dispense: 60 tablets

    Dispensing Fee: $4.75

    Rx Charge: $61.53

    Gross Profit: _____

    Net Profit: _____

2. levofloxacin 500 mg

    Pharmacy's Purchase Price: $250.00 per 30 tablets

    Dispense: 7 tablets

    Dispensing Fee: $3.25

    Rx Charge: $72.39

    Gross Profit: _____

    Net Profit: _____

3. losartan 100 mg

    Pharmacy's Purchase Price: $150.00 per 100 tablets

    Dispense: 60 tablets

    Dispensing Fee: $5.75

    Rx Charge: $120.50

    Gross Profit: _____

    Net Profit: _____

4. Singulair® 4 mg chewable

    Pharmacy's Purchase Price: $230.00 per 90 tablets

    Dispense: 60 tablets

    Dispensing Fee: $4.75

    Rx Charge: $189.67

    Gross Profit: _____

    Net Profit: _____

5. minocycline 100 mg

   Pharmacy's Purchase Price: $110.59 per 60 capsules

   Dispense: 14 capsules

   Dispensing Fee: $3.50

   Rx Charge: $36.79

   Gross Profit: _____

   Net Profit: _____

## Sampling the Certification Exam:

1. Max's Pharmacy has an overhead of $783,385.00. Max wants to make a 30% profit. How much must be sold in goods and services in order to meet this goal?

   a. $235,016.00

   b. $245,040.00

   c. $1,018,401.00

   d. $1,018,420.00

   Answer: _____

2. A pharmacy's overhead is $385,305.00 and income is $532,305.00. What is the percent profit for this pharmacy?

   a. 28%

   b. 30%

   c. 35%

   d. 38%

   Answer: _____

3. A patient is receiving 45 pregabalin 75 mg capsules. The cost for 30 capsules is $45.50. The cost to dispense is $3.75, and the charge to the patient is $89.00. What is the net profit to the pharmacy?

   a. $17.00

   b. $61.50

   c. $68.25

   d. $72.00

   Answer: _____

4. _____ is the difference between the overall cost and selling price.
   a. Stock price
   b. Overhead cost
   c. Gross profit
   d. Net profit

   Answer: _____

5. Actual profit generated is known as
   a. gross profit.
   b. capitation.
   c. turnover.
   d. net profit.

   Answer: _____

## Lesson 25 Content Check

1. A pharmacy determines that its income for the week was $34,780 and the profit was $4,250. What was the overhead for the week?

   Answer: _____

2. A pharmacy has a weekly overhead of $14,500. If this pharmacy has a yearly goal to make a 15% profit, what must it's sales of goods and services amount to each week?

   Answer: _____

3. If a pharmacy has an overhead of $1,909,500, how much income must they make to obtain a 25% profit?

   Answer: _____

4. What is the percent profit of a pharmacy with an income of $1,405,300 and an overhead of $1,244,984?

   Answer: _____

5. How much money is a 15% profit on a pharmacy with an overhead of $2,500,000?

   Answer: _____

## LESSON 25: Income, Overhead, and Profit

6. What is the percent profit of a pharmacy with an income of $1,545,000 and an overhead of $1,237,850?

   Answer: _____

7. How much money is an 8% profit on a pharmacy with an overhead of $1,800,000?

   Answer: _____

8. What is the net profit for:
   Amoxicillin 500 mg caps
   Acquisition cost: $59.65
   Rx price: $72.38
   Dispensing fee: $4.50

   Answer: _____

9. What is the gross profit for:
   Metformin XR 750 mg tabs
   Acquisition cost: $120.99
   Rx price: $149.22
   Dispensing fee: $1.99

   Answer: _____

10. If a pharmacy has an overhead of $798,000, how much income must they make to obtain a 18% profit?

    Answer: _____

11. Use the following expenses to determine your answers to the next few questions:

| Pharmacist salary | $290,000 |
|---|---|
| Technician salary | $150,000 |
| Rent | $65,000 |
| Utilities | $6,800 |
| Computer Maintenance | $2,000 |
| Software subscriptions | $2,250 |
| Liability Insurance | $8,000 |
| Business Insurance | $11,000 |
| Drug purchases | $1,800,980 |

a. What is the total overhead?

Answer: _____

b. How much money is a 25% profit?

Answer: _____

c. To make this goal, how much income must the pharmacy make?

Answer: _____

d. If the pharmacy made $3,876,400.00 what is the percent profit?

Answer: _____

12. Use the following expenses to determine your answers to the next few questions:

| | |
|---|---|
| Pharmacist salary | $180,000 |
| Technician salary | $102,000 |
| Rent | $40,000 |
| Utilities | $4,200 |
| Computer Maintenance | $1,500 |
| Software subscriptions | $2,000 |
| Liability Insurance | $3,000 |
| Business Insurance | $8,000 |
| Drug purchases | $780,000 |

a. What is the total overhead?

Answer: _____

b. How much money is a 5% profit?

Answer: _____

c. To make this goal, how much income must the pharmacy make?

Answer: _____

d. If the pharmacy made $1,090,000.00 what is the percent profit?

Answer: _____

13. The cost of 100 pain tablets is $2.35. What should the selling price be for these tablets in order to yield a 58% percent profit on the cost?

Answer: _____

14. If a retail pharmacy has a total income of $2 million and overhead of $1,750,000, what is their amount of profit?

Answer: _____

# LESSON 26

# Mark Up, Discounts, Depreciation

## Markup:

Markup is the difference between the cost of an item and what it sells for; therefore, markup can be used interchangeably with the word profit. It can be added to any item that a pharmacy, or any business, sells. As discussed in the previous lesson, many terms are used as synonyms of one another so that the following formula can be used to determine the amount of markup:

> Formula:  Markup ($) = Sales Price ($) − Acquisition Cost ($)
> (Synonym)  (Profit)        (Income)           (Overhead)

Once again, there is a difference between markup and percent markup, as it was in the discussion of profit and percent profit.

> Follow up: Therefore, percent markup will always be calculated and compared to the **(sales price / acquisition cost)**, which represents 100%.

> **Example:** A thermometer is sold for $3.89 and it cost the pharmacy $2.52 from the wholesaler. What is the percent markup of this item?

Step 1) Calculate the $ of markup by using the formula above.

$3.89 − $2.52 = $1.32

Sales   Cost   Markup

Step 2) Using the ratio-proportion method, calculate how much % markup this is over the cost (which is 100%)

$$\frac{\$2.52}{100\%} = \frac{\$1.32}{x\%} \quad \text{where } x = \boxed{52.38\ \%}$$

> Remember to NOT use your calculator's % button as in this formula, % is a unit used to make sure things align as they should for the problem to be make logical sense.

**STOP AND PRACTICE:** Determine the following:

1. An asthma tablet costs the pharmacy $24.80 for a month's supply, and it is sold for $30.75. Calculate the markup amount in dollars.

   Answer: _____

2. You gather the following information for famotidine 40 mg tabs:

   Cost: $84.30 for 30 tabs

   Dispense: 30 tabs

   Rx charge: $95.00

   What is the percent markup?

   Answer: _____

3. What is the selling price of an oral antibiotic suspension that costs the pharmacist $15.60 per bottle if there is a 25% markup?

   Answer: _____

4. Antiviral ointment costs a pharmacy $12.50 per tube. The standard markup is 30%. What is the total selling price of a box of 12 tubes?

   Answer: _____

# Discounts:

A discount is a dollar amount or percent of money that is deducted from the total sales price of an item. Usually, pharmacies only deal with discounts on OTC items, but occasionally wholesalers will offer discounts if a certain quantity of items are purchased in a given timeframe. No matter how it is given in a problem, all discount calculations can be solved by using a simple ratio-proportion set up with similar logic to the profit and markup calculations discussed previously.

> **Example:** A box of generic allergy medication is regularly sold for $14.99. If the pharmacy offers a 25% discount for cough and cold season, what would the new selling price be?

Step 1) If acquisition cost = 100 %, then a discounted sales price can be determined by subtracting the percent of the discount from 100%.

   100% – 25%  =  75%
   Cost   Discount   New Sales Price

> A 25% discount means that you are only paying for 75% of the original cost!

Step 2) Using the ratio-proportion method, calculate the selling price of the item.

$$\frac{\$14.99}{100\%} = \frac{\$ x}{75\%} \quad \text{where } x = \boxed{\$ 11.24}$$

> Remember to NOT use your calculator's % button as in this formula, % is a unit used to make sure things align as they should for the problem to be make logical sense.

**STOP AND PRACTICE:** Calculate the selling price of the following discounted items:

1. Cough syrup: regular selling price $5.89 discounted 20%

    Answer: _____

2. Facial tissues: regular selling price $1.19 discounted 15%

    Answer: _____

3. Hair color kit: regular selling price $7.29 discounted 30%

    Answer: _____

4. Body lotion: regular selling price $5.69 discounted 15%

    Answer: _____

5. Baby shampoo: regular selling price $3.89 discounted 25%

    Answer: _____

# Depreciation:

All products lose their value over time, especially with repeated use. Depreciation is a measure of the value lost over a given amount of time, and sometimes is necessary to calculate in pharmacy practice.

Formula: $$\text{Annual Depreciation (\$ / yr)} = \frac{\text{Total Cost (\$)} - \text{Disposal Value (\$)}}{\text{Estimated Life (years)}}$$

> **Example:** A hospital pharmacy bought a sterile hood for $40,350. If it costs $1,200 to be picked up by the appropriate disposal service and it lasted for 13 years, what was the annual depreciation for this hood?

Step 1) Plug the information into the formula to determine the answer.

$$\frac{\$40{,}350 - \$1{,}200}{13 \text{ years}} = \boxed{\$3{,}011.54/\text{yr}}$$

Therefore, this sterile hood's value would depreciate, or fall, by $3,011.54 per year. This amount of money is not technically going anywhere, but it might be an important factor to know if the hospital wanted to try to sell it before it wears out completely and cannot be used again. It is important to understand that depreciation is a theoretical loss in value, not an actual amount of money being lost to any specific entity.

**STOP AND PRACTICE:** Calculate the following:

1. A pharmacy has a new cash register system. The system costs $8294.00 and should last six years. Its disposal value is $890.00. What is the annual depreciation?

   Answer: _____

2. A drug store has a van for home deliveries, and its total cost was $18,452.00. Its disposal value is $2,300.00. It has an estimated life of 4.5 years. What is the annual depreciation?

   Answer: _____

3. A pharmacy has some lab equipment that costs $1478.00. Its disposal value is $150.00 and its estimated life is 5 years. What is the annual depreciation?

   Answer: _____

## Sampling the Certification Exam:

1. Ear drops are priced at $2.59 each. What is the selling price if a 5% discount is offered?

   a. $2.09
   b. $2.46
   c. $2.54
   d. $2.72

   Answer: _____

**LESSON 26:** Mark Up, Discounts, Depreciation

2. A bag of cough drops costs the pharmacy $0.99. The pharmacy sells each bag for $1.69. What is the percent markup?

   a. 71%

   b. 59%

   c. 41%

   d. 36%

   Answer: _____

3. The pharmacy has decided on a 25% markup rate for all OTC antifungal medications. If a tube costs the pharmacy $0.75, what will be the selling price?

   a. $1.10

   b. $1.00

   c. $0.94

   d. $0.89

   Answer: _____

4. The pharmacy needs a new scale. A scale is purchased for $3,999.99 and it is expected to last for 10 years. Its disposal value is $600.00. What is the annual depreciation?

   a. $60.00

   b. $290.00

   c. $340.00

   d. $399.00

   Answer: _____

5. The list price of an antacid suspension is $6.69 per pint, and the discount is 30%. Calculate the net cost of merchandise of this suspension.

   a. $3.89

   b. $4.68

   c. $5.19

   d. $6.39

   Answer: _____

# Lesson 26 Content Check

1. If Allegra D® 60 mg in a 30-pack container costs the pharmacy $32.50, and the price is $55.50, what is the markup amount?

   Answer: _____

2. You just bought a brand-new car for $24,500. If the disposal value of this car is $800 and it should last you 12 years, what is the depreciation rate per year?

   Answer: _____

3. What is the final price on a bottle of Ibuprofen that is originally $7.99 but has been discounted 15%?

   Answer: _____

4. A box of hemorrhoid suppositories costs the pharmacy $8.49 and has a price tag of $11.99 on the OTC shelf. What is the percent markup?

   Answer: _____

5. Eye drops with antihistamine are purchased from the manufacturer in cases of 36 drop-dispenser bottles. The pharmacy desires a markup of $1.75 per bottle. The pharmacy's purchase price is $130.00 per case. What is the selling price per bottle?

   Answer: _____

6. You gather the following information for Tylenol #3 tabs:

   Cost: $32.50 for 500 tabs

   Dispense: 40 tabs

   Rx charge: $3.95

   What is the markup percent?

   Answer: _____

7. The cost of 100 pain tablets is $2.35. What should the selling price be for these tablets in order to yield a 58% percent markup on the cost?

   Answer: _____

8. If a medication costs $93.25 and the selling price is $113.98, what is the markup amount?

Answer: _____

a. What is the percent markup?

Answer: _____

9. A package of a certain drug costs $5.00. The pharmacist applies a 40% markup on cost. What is the cost to the patient?

Answer: _____

10. A hospital pharmacy just purchased two new biological safety cabinets at $18,350.0 each. Each cabinet should last 12 years if maintained properly. The disposal value is $1567.00 each. What is the annual depreciation amount for both cabinets combined?

Answer: _____

11. If a medication costs $43.22 and the selling price is $53.00, what is the markup amount?

Answer: _____

a. What is the percent markup?

Answer: _____

12. A package of a certain drug costs $5.00. The pharmacist applies a 40% markup on cost. What is the cost to the patient?

Answer: _____

13. A drug store has a van for home deliveries, and its total cost was $18,900. If its disposal value is $1,500, and it has an expected lifespan of 7 years, what is its annual depreciation?

Answer: _____

14. What is the selling price of a box of gloves sold at a 40% discount if the original price was $6.99?

Answer: _____

15. What is the sales price of a package of disposable masks if it is bought from the wholesaler for $2.48 and the manager prices all items with a 35% markup?

Answer: _____

16. A bottle of children's vitamins regularly costs $17.99, but is on sale for 20% off. What price will the patient pay at the register?

Answer: _____

17. What is the markup amount on a bottle of St. John's Wort that costs the patient $7.98 and costs the pharmacy $4.30?

Answer: _____

18. A pharmacy has a "15% off" sale going on for all vitamins and herbs. A patient gets 3 bottles of Omega-3 Fish Oil capsules regularly priced at $18.99 each. What will the total be at the register for all 3 bottles?

Answer: _____

# LESSON 27

# Patient Insurance Calculations

## Terminology:

Define the following in order to begin to get an idea of how patient insurance calculations are performed in field of pharmacy –

Premium: _____

_____

Deductible: _____

_____

Co-pay: _____

_____

Formulary: _____

_____

## Understanding the Money:

Review the following graphic to "see" how money (the red arrows) usually moves around in the world of pharmacy:

# UNIT 4: Pharmacy Business Math

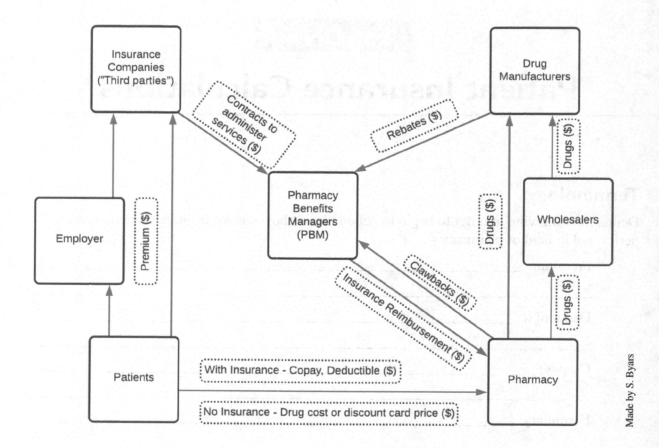

## Discussion:

Patients have healthcare coverage either through their employer or directly from an insurance company, also known as "third parties". No matter which way a patient gets their insurance, all insurance companies require a monthly premium to gain access to the benefits that insurance provides. Note that there are different insurance plans for medical, prescription, vision and dental services, and that this text will only cover information for prescription insurance.

On top of the monthly premium, patients often must pay a yearly deductible before their benefits can kick-in. The amount of a deductible can vary with the plan the patient's chose. Each plan also has a list of covered drugs, or formulary, that is then split into tiers, or groups, that define the copay, or dollar amount, that a patient will pay at a pharmacy for their medication. Tier 1 consists of all generic drugs and is usually the least expensive copay, Tier 2 consists of all the preferred brand-name drugs, and Tier 3 consists of all the non-preferred brand-name drugs and is usually the most expensive copay. During open enrollment, or whenever initially signing up for insurance, a patient can choose the plan that fits their needs best by reviewing the formulary and comparing it to the drugs they are currently taking.

## Private Insurance Calculations:

Use this information from a private insurance company to answer the following questions to better understand how private insurance costs can influence patient's choice of plan:

| *Prescription Saver 4 U* | Drug Plan A | Drug Plan B | Drug Plan C |
|---|---|---|---|
| **Premium** | $200 | $175 | $150 |
| **Deductible** | $500 | $300 | $250 |
| **Generic (Tier 1)** | $10 | $15 | $20 |
| **Brand Preferred (Tier 2)** | $20 | $30 | $35 |
| **Brand Non-preferred (Tier 3)** | $35 | $45 | $50 |

**STOP AND PRACTICE:** Consider the following scenarios and answer the questions that follow.

1. Patient: Abbie Stark    Drug Plan: A    Drug: Crestor® (Tier 2)    Drug Cost: $256.99

   a. How much is the patient's copay?

   Answer: _____

   b. If the patient has already met their deductible, and this is the only prescription the patient gets every month, how much have they paid in total for that prescription (premium + copay)?

   Answer: _____

2. Patient: Carrie Monk    Drug Plan: B    Drug: Crestor® (Tier 2)    Drug Cost: $256.99

   a. How much is the patient's copay?

   Answer: _____

   b. If the patient has already met their deductible, and this is the only prescription the patient gets every month, how much have they paid in total for that prescription (premium + copay)?

   Answer: _____

3. Patient: Jane Lovell    Drug Plan: C    Drug: Crestor® (Tier 3)    Drug Cost: $256.99

   a. How much is the patient's copay?

   Answer: _____

   b. If the patient has already met their deductible, and this is the only prescription the patient gets every month, how much have they paid in total for that prescription (premium + copay)?

   Answer: _____

Follow-up: All three of these patients have benefits through the same insurance company and are on the same drug. What is the difference between them? _____
_____
_____

Follow-up: What does this teach you about selecting plans? _____
_____
_____

Using the same insurance information, answer the following questions:

**STOP AND PRACTICE:** Consider the following scenarios and answer the questions that follow.

1. Patient: Viola Ercole    Drug Plan: C    Drug: lovastatin (Tier 1)    Drug Cost: $21.89

   a. How much is the patient's copay?

   Answer: _____

   b. If the patient has already met their deductible, and this is the only prescription the patient gets every month, how much have they paid in total for that prescription (premium + copay)?

   Answer: _____

2. Patient: Lonnie Smith    Drug Plan: C    Drug: atorvastatin (Tier 1)    Drug Cost: $74.62

   a. How much is the patient's copay?

   Answer: _____

# LESSON 27: Patient Insurance Calculations

b. If the patient has already met their deductible, and this is the only prescription the patient gets every month, how much have they paid in total for that prescription (premium + copay)?

Answer: _____

Follow-up: These patients are getting different drugs but have the same insurance company <u>and</u> the same plan of coverage. What is the difference between the two? _____
_____
_____

Follow-up: What does this teach you about selecting plans? _____
_____
_____

Using the same insurance information, answer the following questions:

**STOP AND PRACTICE:** Consider the following scenarios and answer the questions that follow.

1. You are selecting a new plan during your insurance's open enrollment period. You know that you take the following prescription on a monthly basis:

    diltiazem 120 mg    Drug Cost: $145.22 per month

    You know that your drug is a generic, so it will be in Tier 1 of whichever plan you choose. Work out the following to determine the best plan for you:

    a. Plan A

    i. Monthly cost:

    Answer: _____

    ii. Yearly cost:

    Answer: _____

    b. Plan B

    i. Monthly cost:

    Answer: _____

    ii. Yearly cost:

    Answer: _____

c. Plan C

   i. Monthly cost:

   Answer: _____

   ii. Yearly cost:

   Answer: _____

   Which plan do you choose and why? _____

   _____

## Government (Medicare) Insurance Calculations

Government insurance calculations, especially those related to Medicare, can be more complicated simply because there are more elements to consider. Patients have to not only consider a coinsurance, or percentage-based payment, but also coverage limits that change once a certain price point is exceeded in benefits. A serious point of concern for many Medicare Part D patients is the "donut hole", which is sometimes a staggeringly large gap in coverage where the patient is responsible for a significant portion of their drug costs until they meet the next coverage limit.

Use the following problems to determine a baseline logic that can be applied to many government insurance calculations, or any coinsurance or discount card that offers a percentage of the cost as their baseline payment. Always use the drug cost over 100% as the known variable to compare to.

**STOP AND PRACTICE:** Use the following Medicare part D prescription plan graphic to answer the questions that follow:

| Catastrophic Coverage | Enrollee pays 5%; Plan pays 20%; Medicare pays 75% |
|---|---|
| | Coverage Limit = $7,500 in total drug costs |
| Coverage Gap; aka: the "donut hole" | **Brand-name drugs**: Enrollee pays 40%; Plan pays 10%; 50% is paid by manufacturer discounts |
| | **Generic drugs**: Enrollee pays 75%; Plan pays 25% |
| | Coverage Limit = $3,000 in total drug costs |
| Initial Coverage | Enrollee pays 25%; Plan pays 75% |
| | Deductible = $250 |

1. Patient: Frank Holly    Drug: Benicar®    Cost: $198.02

   a. If the patient has not met their deductible, what will their payment be at the pharmacy?

   Answer: _____

   b. If the patient is in their initial coverage period, what will their payment be at the pharmacy?

   Answer: _____

   i. What amount will the plan cover?

   Answer: _____

   c. If the patient is in the "gap", what will their payment be at the pharmacy?

   Answer: _____

   i. What is the manufacturer's discount?

   Answer: _____

   ii. What amount does the plan pay?

   Answer: _____

   d. If the patient is in the catastrophic coverage period, what will their payment be at the pharmacy?

   Answer: _____

   i. What amount does the plan pay?

   Answer: _____

   ii. What amount does Medicare pay?

   Answer: _____

2. Patient: Loris Benk  Drug: allopurinol  Cost: $82.41

   a. If the patient has not met their deductible, what will their payment be at the pharmacy?

   Answer: _____

   b. If the patient is in their initial coverage period, what will their payment be at the pharmacy?

   Answer: _____

   i. What amount will the plan cover?

   Answer: _____

   c. If the patient is in the "gap", what will their payment be at the pharmacy?

   Answer: _____

   i. What is the manufacturer's discount?

   Answer: _____

   ii. What amount does the plan pay?

   Answer: _____

   d. If the patient is in the catastrophic coverage period, what will their payment be at the pharmacy?

   Answer: _____

   i. What amount does the plan pay?

   Answer: _____

   ii. What amount does Medicare pay?

   Answer: _____

*Consider this scenario*: A patient is prescribed the drug promethazine, which would cost $7.36 for the "cash" price for the quantity they are prescribed. Their copay is $10 at the point of sale. If the patient knew about the drug cost being lower than their co-pay, do you think they would rather pay the cash price? How might situations like this affect how we interact with patients when it comes to dealing with insurance and drug costs?

_____
_____
_____
_____

## Sampling the Certification Exam:

1. Patients need to pay their _____ every month to get access to their drug benefits, or discounts.

    a. Copay

    b. Premium

    c. Deductible

    d. Tier

    Answer: _____

2. Patients have to pay a yearly _____ before their benefits will take effect.

    a. Copay

    b. Premium

    c. Deductible

    d. Formulary

    Answer: _____

3. A specific amount required to be paid by a patient at the pharmacy for a prescription is known as a

    a. benefit.

    b. markup.

    c. copayment.

    d. discount.

    Answer: _____

4. Generic drugs are usually in tier
   a. 1
   b. 2
   c. 3
   d. 4

   Answer: _____

5. A list of approved drugs from an insurance company is known as a
   a. Formulary
   b. Coinsurance
   c. Deductible
   d. Premium

   Answer: _____

## Lesson 27 Content Check

1. Jane Austin is getting a prescription for Crestor®

   The drug costs $259.30 for a one-month supply.

   Jane has a $100 per month drug insurance premium, a $500 deductible and knows that this drug is in Tier 3, for which the copay is $75.

   a. If Jane has not yet met any of her deductible, how much would she pay at the pharmacy for her first fill of Crestor®?

   Answer: _____

   i. How much money has Jane spent in total for her prescription that month (premium + copay)?

   Answer: _____

   b. How many refills of Crestor® would it take to satisfy her deductible?

   Answer: _____

   c. Once she satisfies her deductible, what would her cost be to the pharmacy each month for Crestor®?

   Answer: _____

2. Cooper has a three-month prescription for olmesartan, which would costs $160.09 normally. He has a $75 per month drug insurance premium, a $300 deductible and knows that this drug is in Tier 1, for which the copay is $20.

   a. If he has not yet met any of his deductible, how much would he pay at the pharmacy for the first fill?

   Answer: _____

   i. How much money has Cooper spent in total for his prescription that month (premium + copay)?

   Answer: _____

   b. How many refills of olmesartan would it take to satisfy his deductible?

   Answer: _____

   i. How much time would have elapsed for this to occur?

   Answer: _____

   c. Once he satisfies the deductible, what would the cost be to the pharmacy each month for this prescription?

   Answer: _____

3. Using the graphic from earlier in the lesson, calculate the following Medicare part D drug questions:

   a. A patient who has met their deductible is needing some triamcinolone cream, which costs the pharmacy $82.97 per tube. How much would they pay for the drug if they were still in the initial coverage period?

   Answer: _____

   b. A patient who is in the "gap" approaches the pharmacy. They have a prescription for Vascepa® and request the brand name. If the drug costs the pharmacy $497.87, how much would they pay at the pharmacy?

   Answer: _____

c. A patient in the catastrophic coverage portion of their Medicare coverage plan needs their metformin filled for a 90-day supply. The pharmacy would normally charge $42.99 for this drug. How much will this patient pay for their medication?

Answer: _____

d. A patient who has not yet met their deductible approaches the pharmacy with a prescription for fluconazole, 150 mg #3 tabs. How much would they pay for their prescription at the register if the pharmacy would normally charge $18.99 for this drug?

Answer: _____

e. A Medicare D patient who gets $598.28 in drug costs each month would like to know how long they can expect to be in their initial coverage period. If they have met their deductible, how many months would it take for them to reach the coverage limit?

Answer: _____

f. A patient in the coverage gap of their Medicare D plan is curious to know how many refills of their Humira®, which costs $1,405.13 per month, it would take to reach their catastrophic coverage period. What should the pharmacy technician tell the patient in this situation?

Answer: _____

4. A patient is in their open enrollment period, and has the option to select from the following plans:

| *Drugs Unlimited* | Drug Plan A | Drug Plan B |
|---|---|---|
| Premium | $50 | $50 |
| Deductible | $750 | $300 |
| Generic (Tier 1) | $10 | $15 |
| Brand Preferred (Tier 2) | $25 | $25 |
| Brand Non-preferred (Tier 3) | $35 | $45 |

The patient takes the following:

- Boniva® (Plan A – Tier 3, Plan B – Tier 2); patient gets this drug every month.
- insulin lispro (Plan A and B – Tier 1); patient has to get this drug every 2.5 weeks
- metformin (Plan A and B – Tier 1); patient gets this drug every month.

- Januvia® (Plan A – Tier 2, Plan B – Tier 3); patient gets this drug every month.
- glipidize ER (Plan A and B – Tier 1); patient gets this drug every month.
    a. If the patient chose plan A, what would be their cost each month?

    Answer: _____

    i. What would be their cost over the course of a year?

    Answer: _____

    b. If the patient chose plan B, what would be their cost each month?

    Answer: _____

    i. What would be their cost over the course of a year?

    Answer: _____

    c. Which plan should the patient choose and why?

    Answer: _____

# LESSON 28

# Pharmacy Insurance Calculations

The calculations for different insurance types and how they influence patient payment methods to the pharmacy does not make up the whole picture of where and how money flows through the operations. There are a few more entities that must be discussed to begin to get a general idea of how the finances of a pharmacy operates as a whole.

## Terminology:

Spell out and define the following in order to begin to get an idea of how pharmacy insurance calculations are performed in field of pharmacy –

AWP: _____

_____

U&C: _____

_____

PBM: _____

_____

Once again, review the following graphic to understand the discussion that follows:

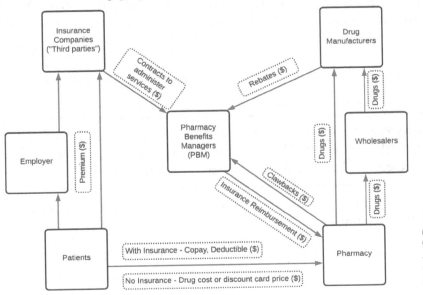

## Discussion:

Pharmacies do not solely rely on the money brought in from the amount a patient pays at the cash register to determine their finances. In fact, most pharmacies rely heavily on reimbursement, or money paid back to the pharmacy from an entity, after adjudicating, or processing, a claim for a patient. In today's practice, most insurance companies do not directly reimburse the pharmacy, nor does the pharmacy directly bill the insurance company – they go through a pharmacy benefits manager (PBM). PBMs are companies who negotiate between drug manufacturers, insurance companies and pharmacies to get the best pricing for all entities involved and uphold these negotiations through contracts. They work directly with the insurance companies and the drug manufacturers, so the pharmacy does not have to, and they are the ones who end up reimbursing the pharmacy.

All drugs have 2 price indicators – AWP, and U&C. AWP, or average wholesale price, is the average worth, or value, of a drug based on information given by a group of entities such as drug manufacturers, insurance contracts, and resources such as The Red Book. U&C, or usual and customary price, is the average worth, or value, of a drug based on the geographic area in which the pharmacy is located, and is typically the price indicator that patients who do not have insurance, often called paying "cash", are charged by. A pharmacy can change the U&C of a drug to try to get a higher reimbursement, but not the AWP. A good analogy would be that AWP is equivalent to the national average price per gallon of gas, and U&C is equivalent to the local average price per gallon of gas. Local entities can bump pricing to help increase profit margins, but the national average will not be affected on the whole.

Typically, a contract with a PBM will stipulate the specific price indicator that they will use for billing purposes and if they allow a dispensing fee to be charged in addition to the amount submitted for reimbursement. The AWP should always be set over 100% as the known to compare using ratio-proportion.

Follow-up: How can pharmacy technicians therefore ensure that they are being reimbursed at the highest rate possible, or in general, ensure they are getting the highest possible profit? _____

_____

_____

_____

> **Example**: What is the retail price of a prescription that has an AWP of $209.84 for a bottle of 500 tabs if a patient is getting a quantity of 60 and the PBM contracted rate is "97% AWP + $1.50 professional fee"?

Step 1) Determine the AWP for the prescription since the quantity doesn't match the size of the container.

$$\frac{\$209.84}{500 \text{ tabs}} = \frac{\$X}{60 \text{ tabs}} \quad \text{where } X = \$25.18$$

# LESSON 28 : Pharmacy Insurance Calculations

Step 2) Determine the rate at which the PBM will allow to be billed, since it's not 100%.

$$\frac{\$25.18}{100\%} = \frac{\$X}{97\%} \quad \text{where } X = \$24.43$$

Step 3) Add the professional fee since it is allowed by the PBM.

$$\$24.43 + \$1.50 = \boxed{\$25.93}$$

**STOP AND PRACTICE:** Calculate the following prices as indicated by the PBM contract formulas that follow.

1) Calculate the retail price of the following prescriptions using the formula "AWP + professional fee = retail price of the Rx". The professional fee is determined by the following chart:

   | AWP for the Quantity of the Rx | Professional Fee |
   |---|---|
   | Less than $20.00 | $4.00 |
   | $20.01–$50.00 | $5.00 |
   | Greater than $50.01 | $6.00 |

   a. verapamil SR tabs #30     AWP/100 = $120.85

   Retail Price: _____

   b. glyburide 5 mg tabs #30     AWP/1000 = $440.05

   Retail Price: _____

   c. dexamethasone 4 mg tabs #12     AWP/100 = $58.40

   Retail Price: _____

   d. acyclovir 200 mg caps #100     AWP/100 = $306.38

   Retail Price: _____

2) Calculate the amount that the pharmacy will submit for reimbursement on each prescription based on the contracted rate of "AWP less 10%". The pharmacy is not permitted to charge a dispensing fee to this PBM for these prescriptions.

   a. bumetanide 2 mg #15 tabs     AWP/90 tabs = $48.90

   Amount of Claim: _____

b. furosemide 20 mg #30 tabs     AWP/100 tabs = $84.07

Amount of Claim: _____

c. hydrochlorothiazide 12.5 mg #60 caps    AWP/1000 caps = $30.25

Amount of Claim: _____

Sometimes, contracts will include room for both price indicators, but will only reimburse the pharmacy for the least amount of money that is billed. In these cases, both prices will need to be calculated and then compared to see which is least to determine what the PBM would allow. PBMs also can set maximum thresholds that cannot be exceeded when billing, often referred to as the maximum allowable cost (MAC).

> **Example**: If the pharmacy has an agreement with a PBM for reimbursement of "90% AWP or 100% U&C (whichever is less) + a $3.50 dispensing fee", what is the total amount of the claim that will be submitted for #30 benzonatate 100 mg capsules (AWP/100 is $99.42 and U&C/100 is $103.08)?

Step 1) Determine the AWP and U&C for the prescription since the quantity doesn't match the size of the container.

$$\text{AWP: } \frac{\$99.42}{100 \text{ caps}} = \frac{\$X}{30 \text{ caps}} \quad \text{where } X = \$29.83$$

$$\text{U\&C: } \frac{\$103.08}{100 \text{ caps}} = \frac{\$X}{30 \text{ caps}} \quad \text{where } X = \$30.92$$

Step 2) Determine the rate at which the PBM will allow to be billed, since the AWP is not at 100%.

$$\text{AWP: } \frac{\$29.83}{100\%} = \frac{\$X}{90\%} \quad \text{where } X = \$26.85$$

U&C is allowed to be billed at 100% so the price remains the same at $30.92

Step 3) Compare the two prices, and add the dispensing fee to the lowest one (as stipulated in the contract)

AWP at $26.85 is less than U&C at $30.92, so the final calculation would be:

$$\$26.85 + \$3.50 = \boxed{\$30.35}$$

# LESSON 28 : Pharmacy Insurance Calculations

**STOP AND PRACTICE:** Calculate the following prices as indicated by the PBM contract formulas that follow.

1. If the pharmacy has an agreement with a third-party plan for reimbursement of "87% AWP or 100% U&C (whichever is less) + a $3.50 dispensing fee", what is the total amount of the third-party claim for the following prescriptions?

    a. verapamil SR tabs #30    AWP/100 = $120.85    U&C/30 = $29.99

    Amount of Claim: _____

    b. acyclovir 200 mg caps #100    AWP/100 = $306.38    U&C/30 = $95.39

    Amount of Claim: _____

    c. doxepin 150 mg caps #30    AWP/100 = $66.50    U&C/30 = $28.60

    Amount of Claim: _____

    d. tetracycline 250 mg caps #28    AWP/1000 = $52.43    U&C/28 = $1.17

    Amount of Claim: _____

## Sampling the Certification Exam:

1. The national average cost of a drug product to a pharmacy is known as the

    a. cost-benefit analysis.
    b. net cost.
    c. markup.
    d. average wholesale price.

    Answer: _____

2. The AWP for 100 tablets is $56.98. The pharmacy is able to order this medication at AWP minus 7%. What is the cost to the pharmacy for 60 tablets?

    a. $31.79
    b. $32.62
    c. $34.19
    d. $37.87

    Answer: _____

3. The AWP for 450 mL of a medication is $79.85. The pharmacy charges AWP plus 5.25% plus a $3.99 dispensing fee per prescription. What will the patient be charged for a prescription of 360 mL?

   a. $64.80
   b. $68.20
   c. $71.22
   d. $72.40

   Answer: _____

4. A third-party provider reimburses a pharmacy AWP less 15% plus a professional fee of $4.00. The prescription in question is for 50 capsules having an AWP of $25 per 100 capsules. Calculate the amount that the pharmacy will be reimbursed.

   a. $1.88
   b. $10.62
   c. $12.50
   d. $14.63

   Answer: _____

5. Pharmacists can greatly help reduce patients' drug expenses by making them aware of
   a. herbal remedies.
   b. over-the-counter drugs.
   c. comparable, lower-priced drugs.
   d. brand name drugs.

   Answer: _____

## Lesson 28 Content Check

1. The pharmacy purchases 100 sulfamethoxazole/trimethoprim tablets for $71.35. The PBM will reimburse the pharmacy using the formula "AWP + 3.5% + $4.50 dispensing fee".

   a. If the pharmacy dispenses a prescription of 50 tablets, what is the pharmacy's cost for the prescription?

   Answer: _____

b.  If the AWP for this drug is $103.88 for 100 tablets, what is the total amount that the pharmacy will submit to the insurance company for reimbursement on this prescription?

Answer: _____

c.  How much profit will the pharmacy make on this prescription?

Answer: _____

2.  The pharmacy purchases 5 metered-dose inhalers at a cost of "AWP less 3%". The third-party payer will reimburse with the formula of "AWP + 5%" and does not allow a dispensing fee. If the AWP for each inhaler is $36.35, and the pharmacy dispenses two inhalers, what is the pharmacy's cost of the prescription?

Answer: _____

3.  Calculate the amount that the pharmacy will submit for reimbursement on each prescription based on the formula "AWP plus 4%". The pharmacy charges a $6.25 dispensing fee for each prescription.

   a.  Terazosin capsule #30     AWP/500 caps = $120.68

   Amount of Claim: _____

   b.  Sotalol tablets #60     AWP/100 tabs = $39.78

   Amount of Claim: _____

   c.  Cozaar® tablets #15     AWP/90 tabs = $317.50

   Answer: _____

4.  The AWP for 500 tablets is $130.29. The pharmacy is able to order this medication at AWP minus 13%. What is the cost to the pharmacy for 120 tablets?

Answer: _____

5.  The AWP for 60 tablets is $65.98. The pharmacy charges AWP plus 4.5% plus a $5.50 dispensing fee per prescription. What will the patient be charged for 90 tablets?

Answer: _____

6. The AWP for 480 mL of a medication is $63.20. The pharmacy charges AWP plus 3% plus a $4.75 dispensing fee per prescription. What will the patient be charged for a prescription of 180 mL?

Answer: _____

7. Determine the selling price and pharmacy profit of the following prescriptions using the following charts. Assume all patients have the same insurance. Remember that patients without insurance usually pay by the U&C price indicator.

| Drug | Strength | Package Size | Acquisition Cost | U&C | AWP | Tier |
|---|---|---|---|---|---|---|
| Amoxicillin | 400 mg | 20 | $18.43 | $24.94 | $32.65 | 1 |
| Cytomel | 5 mcg | 100 | $82.95 | $84.10 | $79.16 | 2 |
| Zanaflex | 2 mg | 90 | $238.63 | $304.77 | $337.45 | 3 |

| U&C/AWP | Dispensing Fee |
|---|---|
| $0 - $49.99 | $2.00 |
| $50.00 - $99.99 | $4.00 |
| $100.00 + | $6.00 |

| Tier | CoPay's |
|---|---|
| 1 | $10.00 |
| 2 | $35.00 |
| 3 | $55.00 |

a. Mrs. Jones needs #10 Amoxicillin 400 mg, but does not have insurance.

Patient Pays: _____

Pharmacy Profits: _____

b. Mr. Warren needs #40 Amoxicillin 400 mg, and has insurance. They reimburse the pharmacy at AWP less 10%, or 100% U&C, whichever is less, plus a dispensing fee for this drug.

Patient Pays: _____

Amount to submit for reimbursement: _____

Pharmacy Profits: _____

c. Mrs. Goose needs #30 Cytomel 5 mcg and does not have insurance.

Patient Pays: _____

Pharmacy Profits: _____

d. Mrs. Goose's husband also needs #30 Cytomel 5 mcg, but he does have insurance through his employer. They reimburse the pharmacy at AWP + 5%, or 100% U&C, whichever is less, plus a dispensing fee for this drug.

Patient Pays: _____

Amount to submit for reimbursement: _____

Pharmacy Profits: _____

e. Mr. Blum needs #30 Zanaflex 2 mg and does not have insurance.

Patient Pays: _____

Pharmacy Profits: _____

f. Mrs. Whisk needs #30 Zanaflex 2 mg and has insurance. She also has a COB coupon card from the manufacturer that allows her to pay 60% of her copay. Her insurance company reimburses the pharmacy at AWP less 12%, or 100% U&C, whichever is less, for this drug.

Patient Pays: _____

Amount to submit for reimbursement: _____

Pharmacy Profits: _____

## Unit 4 Content Review

**Multiple Choice** - *Identify the choice that best completes the statement or answers the question.*

1. Julie's Pharmacy has a monthly overhead of $16,890.00. If the owner wants to make a 20% profit, what must the pharmacy's income be?

Answer: _____

2. If a pharmacy's overhead is $759,389.00 and the income is $938,223.00, what is the percent profit for this pharmacy?

Answer: _____

3. From the following information, calculate the gross profit.

   Prinivil® 10 mg; Dispense: 30 tablets

   Pharmacy's Purchase Price: $43.98 per 100 tablets

   Dispensing Fee: $4.00

   Rx Charge: $35.89

   Answer: _____

4. From the following information, calculate the net profit.

   metformin 850 mg; Dispense: 60 tablets

   Pharmacy's Purchase Price: $37.99 per 100 tablets

   Dispensing Fee: $4.75

   Rx Charge: $61.53

   Answer: _____

5. A cough syrup normally sells for $3.59. The pharmacy is offering a 15% discount. What is the new selling price?

   Answer: _____

6. Ear drops are priced at $2.59 each. What is the selling price if a 5% discount is offered?

   Answer: _____

7. A tube of hydrocortisone cream costs the pharmacy $1.99. This tube is sold to the patient for $2.95. What is the markup rate?

   Answer: _____

8. The AWP for 100 tablets is $56.98. The pharmacy is able to order this medication at AWP minus 7%. What is the cost to the pharmacy for 60 tablets?

   Answer: _____

9. Cefaclor 250 mg capsules can be ordered in a bottle of 100 capsules. The maximum inventory level is 250 capsules and the minimum level is 50 capsules. If the current inventory is 25 capsules, how many bottles need to be ordered to reach the minimum?

   Answer: _____

## LESSON 28 : Pharmacy Insurance Calculations

10. Fluoxetine 20 mg capsules can be ordered in a bottle of 90 capsules. The maximum inventory level is 240 capsules and the minimum level is 60 capsules. If the current inventory is 25 capsules, how many bottles need to be ordered to reach, but not exceed, the maximum?

    Answer: _____

11. The pharmacy buys a new delivery car for $17,850.00. They expect the car to last 6 years and its disposal value is $2,500.00. What is the annual depreciation?

    Answer: _____

12. The pharmacy needs a new scale. A scale is purchased for $3,999.99 and it is expected to last for 10 years. Its disposal value is $600.00. What is the annual depreciation?

    Answer: _____

13. Lilly's Pharmacy has an overhead of $783,385.00. Lilly wants to make a 30% profit. How much must be sold in goods and services in order to meet this goal?

    Answer: _____

14. A pharmacy's overhead is $385,305.00 and income is $532,305.00. What is the percent profit for this pharmacy?

    Answer: _____

15. From the following information, calculate the net profit.

    Levaquin® 500 mg; Dispense: 7 tablets

    Pharmacy's Purchase Price: $250.00 per 30 tablets

    Dispensing Fee: $3.25

    Rx Charge: $72.39

    Answer: _____

16. From the following information, calculate the gross profit.

    oxcarbazepine 600 mg; Dispense: 60 tablets

    Pharmacy's Purchase Price: $150.00 per 100 tablets

    Dispensing Fee: $5.75

    Rx Charge: $120.50

    Answer: _____

17. The pharmacy is having a sale on OTC analgesics. If ibuprofen tablets are originally priced at $1.99, what is the selling price after a 20% discount?

Answer: _____

18. A 50% discount is being offered on all lip balm in honor of the Fall changes If the normal price is $2.25, what is the discounted price?

Answer: _____

19. The pharmacy has decided on a 25% markup rate for all OTC antifungal medications. If a tube costs the pharmacy $0.75, what will be the selling price?

Answer: _____

20. The markup for an allergy tablet is 15%. This tablet costs the pharmacy $98.70 for a bottle of 100 tablets. What is the selling price for a prescription of 30 tablets?

Answer: _____

21. The AWP for 500 tablets is $130.29. The pharmacy is able to order this medication at AWP minus 13%. What is the cost to the pharmacy for 120 tablets?

Answer: _____

22. The AWP for 250 capsules is $259.65. If the pharmacy dispenses 45 capsules at AWP plus 3% plus a $3.75 dispensing fee, what would the cost to the patient be?

Answer: _____

23. The AWP for 480 mL of a medication is $63.20. The pharmacy charges AWP plus 3% plus a $4.75 dispensing fee per prescription. What will the patient be charged for a prescription of 180 mL?

Answer: _____

24. The AWP for 450 mL of a medication is $79.85. The pharmacy charges AWP plus 5.25% plus a $3.99 dispensing fee per prescription. What will the patient be charged for a prescription of 360 mL?

Answer: _____

LESSON 28 : Pharmacy Insurance Calculations

25. Amoxicillin 500 mg capsules can be ordered in a bottle of 500 capsules. The maximum inventory level is 550 capsules and the minimum level is 150 capsules. If your current inventory is 200 capsules, how many bottles need to be ordered to reach the minimum?

Answer: _____

26. Minocycline 50 mg capsules can be ordered in bottles of 50 capsules. The maximum inventory level is 200 capsules and the minimum level is 30 capsules. If the current inventory is 27 capsules, how many bottles need to be ordered to reach, but not exceed, the maximum?

Answer: _____

27. Lisinopril 10 mg tablets can be ordered in bottles of 100 tablets. The maximum inventory level is 260 tablets and the minimum level is 60 tablets. Currently, the inventory is at 20 tablets and there are two prescriptions pending for 60 tablets each. How many bottles need to be ordered to reach the minimum?

Answer: _____

28. A pharmacy's average inventory for the last year was $210,032.00 and the annual cost total was $1,783,003.00. What was the turnover rate?

Answer: _____

29. If a pharmacy's average inventory is $987.50 and the annual purchases are $7,538.00, what is the turnover rate?

Answer: _____

30. If a pharmacy's average inventory is $673.00 and the annual purchases are $8,325.00, what is the turnover rate?

Answer: _____

31. A new IV hood is bought by the pharmacy for $7,899.00. It is expected to be used for 8 years. The disposal value is $500.00. What is the annual depreciation?

Answer: _____

32. Ten tubes of ointment are purchased at $5.11 each. If the account is fully paid within 30 days, the manufacturer offers a discount of 20%. What is the total discounted purchase price?

Answer: _____

33. A pharmacy computer system depreciates in value $1,000 each year. If the system was purchased new for $10,000 in 1998 and is being disposed of in 2005, what is its current disposal value?

Answer: _____

# UNIT 5

# Solution Calculations

# LESSON 29

# Understanding Solutions

There are many types of liquids used in the healthcare field, and in the practice of pharmacy, the term is used as a general definition that covers the many variations that exist. One of the most important types of liquids is a solution, but other examples are suspensions, emulsions, tinctures, elixirs, syrups,

## What is a Solution?

A solution is a mixture in which the solute is evenly distributed (i.e., completely dissolved) throughout the solvent. This definition, though extremely accurate and concise, brings up the need for two other definitions. Define them below:

Solute – _____
_____

    The solute in all pharmacy math problems is the _____.

Solvent – _____
_____

    The solvent, or diluent, in **almost** all pharmacy math problems is _____.

    Sometimes, solvents are referred to as a drug's _____ because it drives the drug into the body.

### Solutions

The solution is formed when the solute dissolves in a solvent.

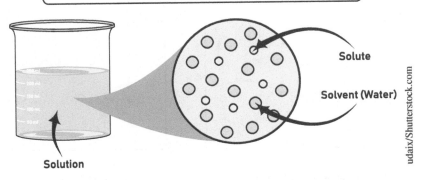

UNIT 5: Solution Calculations

Pharmacy technicians are often tasked with creating solutions for patient use by means of diluting, or adding a diluent to, a previously existing concentrated stock solution. Pharmacies utilize stock solutions to save space on their shelves as they are highly concentrated, low volume mixtures that can be easily stored for future use. They are not to be given out to patients as they are usually too strong, which is why pharmacy technicians must create weaker versions of the same mixture using what is available on the shelf. Concentration itself is a mathematical expression that measures of the amount of solute within a solution or mixture. The process of diluting a solution requires a knowledge of how the process works conceptually as well as the calculations involved. Review the graphic below to get a birds-eye view of the difference between a dilute solution and a concentrated solution.

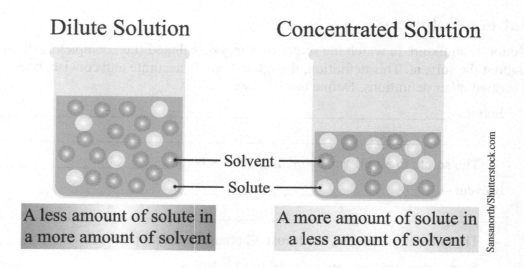

Consider the following image. Assume that each dot represents 1 g of drug:

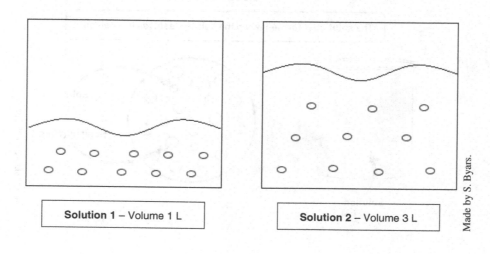

1. How much drug is in solution #1? _____ g

   a. What is this amount of drug referred to as? (Hint: vocabulary word)
      _____

2. What is the volume of solution #1? _____ mL

   b. What is the volume often referred to as? (Hint: vocabulary word)
      _____

3. How much drug is in solution #2? _____ g

4. What is the volume of solution #2? _____ mL

5. Determine the concentration of each solution by using creating a fraction of grams over milliliters, and then reducing them using a proportion set equal to $X$ g/1 mL. Therefore:

   a. What is the ratio concentration of solution #1?

   Answer: _____

   b. What is the ratio concentration of solution #2?

   Answer: _____

   c. **T / F:** The concentration in the two beakers is the same.

6. **T / F:** The amount of drug in the two beakers is the same.

7. **T / F:** The process of dilution alters the amount of drug in the solution.

## Solution Concentration - Expression and Evaluation.

Before solving math problems that concern concentration, it is important to be able to read it and understand it. Often, mathematical values will be given in different "languages" in the problem – we need to make sure the "language", or in this case, the unit, is the SAME. A math problem **cannot** solved unless the units within it are aligned in the correct way.

1. The mathematical value of concentration of a solution/mixture represents a measure of
   _____
   _____

   a. **T / F:** Therefore, the identity or active ingredient of a solution/mixture defines its name.

2. Solution concentrations are often expressed in terms of: _____
   _____

> **Hint:** this is how you will see them expressed in a math problem. It will help you identify them!

Remember:

> To convert from a percent to a ratio:

> **Example:** What is the correct representation of 4% as a ratio?

Step 1) Convert the percent to a fraction of the percent over 100 (because percent means "out of 100").

$$\frac{4}{100}$$

Step 2) Set this equal to $\frac{1}{X}$, solving, and reporting the answer as 1: $X$

$$\frac{4}{100} = \frac{1}{X} \text{ where } X = 20 \text{ so the answer is } \boxed{1 : 20}$$

To convert from a ratio to a percent:

> **Example:** Convert 3:20 to a percent

Step 1) Convert to a fraction.

$$\frac{3}{20}$$

Step 2) Set up a proportion equal to $\frac{X}{100}$. Cross multiply and divide. Add a % sign to the answer you get for $X$.

$$\frac{3}{20} = \frac{X}{100} \text{ where } x = 15, \text{ so the answer is } \boxed{15\%}$$

**STOP AND PRACTICE:** Convert the following:

1. 1:400 -> %

   Answer: _____

2. 25% -> ratio

   Answer: _____

3. 1:2000 -> %

   Answer: _____

**LESSON 29:** Understanding Solutions

4. 3% -> ratio

Answer: _____

5. 0.9% -> ratio

Answer: _____

**STOP AND PRACTICE:** Arrange the following solutions of saline-in-water from least concentrated to most concentrated:

1. Using Ratio Strength:
   a. 1:20
   b. 1:40
   c. 1:5
   d. 1:200    Order: _____
   e. 1:10    *(least)*                                    *(most)*
   f. 1:500

2. Using Percent:
   g. 5%
   h. 2.5%
   i. 20%
   j. 0.5%    Order: _____
   k. 10%    *(least)*                                    *(most)*
   l. 0.2 %

3. Which was easier?
   _____

## Applying Concepts:

1. What is really happening when someone leaves their drink out, the ice melts and then they complain about the taste being "watered down"?
   _____
   _____

2. Based on what you've learned so far, circle the correct word to complete the statement:
   a. As you add solvent, a solution becomes more (**concentrated / diluted**).
   b. By diluting a stock solution, the total volume of the solution (**increases / decreases**).
   c. A diluted solution generally has (**more / less**) volume than a concentrated solution and (**more / less / the same**) amount of solute than a concentrated solution.
   d. Dilution is a process of (**increasing / decreasing**) the concentration of a stock solution by means of adding a diluent.
   e. A stock solution is (**more / less**) concentrated than a diluted solution.
   f. A stock solution is (**stronger / weaker**) than a diluted solution.
3. **T / F:** The diluent used to dilute a stock solution is <u>always</u> the same as the solvent.
   a. Explain.

   _____
   _____

4. **T / F:** You will only be doing the math related to <u>**diluting**</u> a solution, not making it more concentrated.
   a. Let's think: can you make a solution more concentrated? **Y / N**
      i. How?

      _____
      _____

      ii. Will you ever do this in a real pharmacy? **Y / N**

## Sampling the Certification Exam:
1. Express 4:21 as a percent.
   a. 0.2%
   b. 5.3%
   c. 19%
   d. 53%

   Answer: _____

2. What is the fraction represented by the ratio 1:20?
   a. 1/20
   b. 20/1
   c. 1/2
   d. 1/0.2

   Answer: _____

3. What is the percent strength of a 1:100 solution?
   a. 0.1%
   b. 1%
   c. 10%
   d. 100%

   Answer: _____

4. What is the fraction represented by the ratio 1:400?
   a. 1/400
   b. 400/1
   c. 1/4
   d. 1/40

   Answer: _____

5. What is the percent strength of a 1:5 solution?
   a. 20%
   b. 40%
   c. 500%
   d. 10%

   Answer: _____

## Lesson 29 Content Check:

1. What is the solute in the following solution: sodium hypochlorite 2%

   Answer: _____

2. What is the active ingredient in the following mixture: hydrocortisone 2.5%

   Answer: _____

3. What is the solute in the following solution: ethyl alcohol 35%

   Answer: _____

4. What is the active ingredient in the following mixture: triamcinolone 0.0025%

   Answer: _____

5. Assuming both solutions contain isopropyl alcohol as their main ingredient, which solution would be the most concentrated - 1:400 or 5%?

   Answer: _____

6. Assuming both solutions contain insulin as their main ingredient, which solution would be the most dilute - 1:20 or 10%?

   Answer: _____

7. Assuming both solutions contain bleach as their main ingredient, which solution would be the most concentrated - 1:200 or 0.7%?

   Answer: _____

8. Assuming both solutions contain magnesium sulfate as their main ingredient, which solution would be the most dilute - 1:40 or 3%?

   Answer: _____

9. Concentration can be expressed as what type of mathematical relationships?

   Answer: _____

10. Increasing the amount of diluent in a mixture _____ the total concentration of the mixture.

11. The only way to increase the concentration of a solution is to _____.

12. What is the percent strength of a 1:5000 solution?

    Answer: _____

13. Convert 1:4000 to a percent.

    Answer: _____

14. Convert 0.03% to a ratio.

   Answer: _____

15. Convert 1:10 to a percent.

   Answer: _____

16. Convert 1:75 to a percent.

   Answer: _____

17. Convert 0.67% to a ratio.

   Answer: _____

18. Convert 1:300 to a percent.

   Answer: _____

19. Convert 1:50 to a fraction.

   Answer: _____

20. Convert 2/21 to a ratio.

   Answer: _____

# LESSON 30

# Concentration Expression

As discussed in the previous lesson, concentrations are expressed as either a percent or a ratio. There is more meaning than just the numbers though – both expressions represent the amount of active drug over the total solution as follows:

$$X\% \rightarrow \frac{x(\text{active drug})}{100(\text{total solution})} \quad \text{OR} \quad 1:X \rightarrow \frac{1(\text{active drug})}{X(\text{total solution})}$$

Additionally, the concentration of the solution/mixture will always have a letter-based fraction following the numbers that will help you determine the <u>units</u> of the mixture you are dealing with (and a little bit about its identity). There are three types. Explore each below.

## Weight per Volume

1. The unit "w/v" means: _____ and is expressed as: ____/____

   a. This means that the active drug is in a (**solid / liquid**) form, and the total mixture is a (**solid / liquid**).

2. **T / F:** Adding a drug in the form of a powder or crystal to a solution will change the solutions <u>volume.</u>

3. **T / F:** Adding a drug in the form of a powder or crystal to a solution will change the solutions <u>concentration.</u>

   a. How/why? _____
   _____
   _____
   _____

## Volume per Volume

1. The unit "v/v" means: _____ and is expressed as: ____/____

   a. This means that the active drug is in a (**solid / liquid**) form, and the total mixture is a (**solid / liquid**).

2. **T / F**: Adding a drug in the form of a liquid to a solution will change the solutions <u>volume</u>.

   a. How/why? _____
   _____
   _____

   b. Draw a picture:

3. **T / F**: Adding a drug in the form of a liquid to a solution will change the solutions <u>concentration</u>.

   a. How/why? _____
   _____
   _____

## Weight per Weight

1. The unit "w/w" means: _____ and is expressed as: ___/___

   a. This means that the active drug is in a (**solid / liquid**) form, and the total mixture is a (**solid / liquid**).

2. **T / F**: Adding a drug in the form of a solid to a solid mixture will change its <u>weight</u>.

   a. How/why? _____
   _____
   _____
   _____

   b. Draw a picture:

3. **T / F**: Adding a drug in the form of a solid to a solid mixture will change its <u>concentration</u>.

   a. How/why? _____
   _____
   _____

# LESSON 30: Concentration Expression

## Solving Simple Percent Strength Problems

Percent solution problems can be solved by using the ratio-proportion method. The most common mistakes with these types of problems involve not paying attention to the units or forgetting the basics of all the mathematical principles that have been discussed up to this point.

> **Example:** 1:500 w/v solution contains _____ mcg of drug per 100 mL of solution.

Step 1) Write out the definition of the ratio as a fraction with units.

$$\frac{1\,g}{500\,mL}$$

Step 2) Set up a proportion and plug in the unknown.

$$\frac{1\,g}{500\,mL} = \frac{X\,g}{100\,mL} \quad \text{where } X = 0.2\,g$$

Step 3) Since the answer does not match what the question is asking for, convert using the KSMM method.

> 200,000 mcg

**STOP AND PRACTICE:** Define the following concentrations of solutions as a fraction with units, then solve the questions that follow:

1. 43% w/v = _____
   a. How many g are in 50 mL of this solution?
   
   Answer: _____
   
   b. How many mL of solution would 258 g of the active ingredient make?
   
   Answer: _____

2. 0.9% w/v = _____
   a. How many mg are in 900 mL of this solution?
   
   Answer: _____
   
   b. How many mL of solution would 4.5 g of the active ingredient make?
   
   Answer: _____

3. 1:400 w/v = _____
   a. How many g are in 1,500 mL of this solution?
   
   Answer: _____

b. How many L of solution would 0.9 kg of the active ingredient make?

Answer: _____

4. 1:50 w/v = [ _____ ]

   a. How many mcg are in 5 mL of this solution?

   Answer: _____

   b. How many L of solution would 20 g of the active ingredient make?

   Answer: _____

5. 1:2,000 w/v = [ _____ ]

   a. How many kg are in 32 L of this solution?

   Answer: _____

   b. How many L of solution would 320 mg of the active ingredient make?

   Answer: _____

Occasionally, a problem will ask for the percent strength of a mixture by giving the amount of active ingredient and the amount of the total mixture.

> **Example:** What is the percent strength of a 500 mL mixture containing 19 g of active ingredient?

Step 1) Remember: percent means **out of 100**, so always make sure your units make sense with each definition of percent strength (**w/v** should be g/mL, **v/v** should be mL/mL, **w/w** should be g/g). If the units don't match the definition, convert them!

The unit of g and mL tells us that this is a w/v mixture. Those are the correct units for the definition of a percent solution, so there is no need to convert further.

$$\frac{19\,g}{500\,mL}$$

Step 2) Set up equal to a proportion of $X$ over 100 and solve! Don't forget the percent symbol in the answer.

$$\frac{19\,g}{500\,mL} = \frac{X}{100} \text{ where } X = \boxed{3.8\%}$$

> This is one of the few examples where the units on the right side of the equation don't quite make sense according to the ratio proportion method, but again, this set up is relying on the concept that $X/100$ is the definition of percent!

## Stop and Practice:

1. What is the percent strength of a 70 g mixture that contains 45 g of active ingredient?

   Answer: _____

2. What is the percent strength of a half-liter solution containing 29 g of acetaminophen?

   Answer: _____

3. What is the percent strength of a mixture that is 90 mL of water mixed with 25 mL of alcohol?

   Answer: _____

4. What is the percent strength of a solution that contains 450 mg in 5 mL?

   Answer: _____

5. What is the percent strength of a 2 L solution that has 350 mL of active ingredient in it?

   Answer: _____

## Sampling the Certification Exam:

1. If there are 40 g of dextrose in 500 mL of water, what is the percentage strength of the solution?

   a. 200% w/v
   b. 100% w/v
   c. 10% w/v
   d. 8% w/v

   Answer: _____

2. How many milligrams of drug are in 45 mL of a 1.5% w/v solution?

   a. 675 mg
   b. 68 mg
   c. 3 mg
   d. 0.7 mg

   Answer: _____

# UNIT 5: Solution Calculations

3. A 1:250 w/v solution has _____ of active ingredient and _____ of total product.
   a. 1000 g; 250 mL
   b. 0.001 g; 250 mL
   c. 250 g; 1 mL
   d. 1 g; 250 mL

   Answer: _____

4. How many grams of active ingredient are in 500 mL of a 1:300 w/v solution?
   a. 150,000 g
   b. 1500 g
   c. 1.7 g
   d. 0.6 g

   Answer: _____

5. How many milligrams are in 30 mL of a 1:150 w/v solution?
   a. 0.2 mg
   b. 5 mg
   c. 6.7 mg
   d. 200 mg

   Answer: _____

## Lesson 30 Content Check

1. How many grams of drug are in 100 g of a 25% w/w mixture?

   Answer: _____

2. Write out the definition of the following with appropriate units: 20% w/v

   Answer: _____

3. Write out the definition of the following with appropriate units: 15% w/w

   Answer: _____

4. Write out the definition of the following with appropriate units: 8% v/v

   Answer: _____

# LESSON 30: Concentration Expression

5. Write out the definition of the following with appropriate units: 0.9% w/w

    Answer: _____

6. Write out the definition of the following with appropriate units: 1.5% w/v

    Answer: _____

7. Write out the definition of the following with appropriate units: 2% v/v

    Answer: _____

8. Convert a 1:400 w/v solution into a fraction with appropriate units.

    Answer: _____

9. Write out the definition of the following with appropriate units: 1:200 w/v

    Answer: _____

10. Write out the definition of the following as a ratio with appropriate units: 60% w/v

    Answer: _____

11. What is the percent solution (w/v) of a 300 mL solution containing 40g of active drug?

    Answer: _____

12. How many mL of active drug are in 20 mL of a 40% v/v solution?

    Answer: _____

13. How many mL of a 10% w/v dextrose solution can be made by using 50g of dextrose?

    Answer: _____

14. 1.5% w/v =
    a. How many mcg are in 0.5 L of this solution?

    Answer: _____

    b. How many L of solution would 2 kg of the active ingredient make?

    Answer: _____

15. 0.0025% w/v =

   a. How many kg are in 2,500 L of this solution?

   Answer: _____

   b. How many L of solution would 400 mg of the active ingredient make?

   Answer: _____

16. 9% w/v =

   a. How many g are in 25 mL of this solution?

   Answer: _____

   b. How many mL of solution would 81 g of the active ingredient make?

   Answer: _____

17. 50% w/v =

   a. How many g are in 75 mL of this solution?

   Answer: _____

   b. How many mL of solution would 165 g of the active ingredient make

   Answer: _____

18. How many grams of active drug are in 45 mL of a 1:4 w/v solution?

   Answer: _____

19. How many mg of active drug are in 35 mL of a 1:1000 w/v solution?

   Answer: _____

20. If there are 5 g of zinc in 75 g of ointment, what is the percentage strength of the drug in the ointment?

   Answer: _____

## LESSON 31

# Concentration Expression: Identities and Dilution Factors

## Understanding Volume to Volume and Weight to Weight Concentration Expressions:

Essentially all percent strength and/or ratio concentration expressions represent the following logic:

> $X + Y = Z$
> Where:
> $X$ = active drug/solute
> $Y$ = inactive/solvent drug
> $Z$ = total solution

Additionally, many v/v or w/w solutions/mixtures will use ratio's as the mathematical language to express the relationships of ingredients. Remember that ratios are simply comparisons of one value to another. The colon in a ratio can be read as **to** or **in**. These words within a math problem can tell you how much of, and the identity of, each ingredient.

> **Example:** A 1:100 nitroglycerin-**to**-sodium chloride solution can be thought of as 1 part of nitroglycerin **to** 100 parts of sodium chloride.
>
> **Example:** A 1:2,000 bleach-**in**-water solution can be thought of as 1 part bleach **in** 2,000 total parts

1. What is the difference between these two examples? _____

   _____

   Therefore:

   a. The word **to** is comparing the number of parts of _____ to the number of parts of _____

      i. How would you find the total number of parts in this example?

         _____

295

b. The word **in** is comparing the number of parts of _____ to the number of parts of _____

   i. How would you find the inactive/diluent number of parts in this example?
   _____

## Stop and Practice:

1. A solution of **ammonia-to-water** (v/v) is 1:9.

   a. How many parts of ammonia are there?

   Answer: _____

   b. How many parts of water are there?

   Answer: _____

   c. How many total parts are there?

   Answer: _____

2. A solution of **ammonia-in-water** (v/v) is 1:9.

   a. How many parts of ammonia are there?

   Answer: _____

   b. How many parts of water are there?

   Answer: _____

   c. How many total parts are there?

   Answer: _____

3. For a 3-in-18 dilution, what is the ratio of ammonia-to-water?

   Answer: _____

## Converting "Parts" Into Measurable Quantities

The above examples discuss how the language of some math problems may give you more information than meets the eye in any given problem; however, there is no piece of equipment that can measure "parts". We need to be able to use the information above to help us not only solve mathematical problems on paper, but also actually apply them to pharmacy practice.

1. Use the following problems to establish your pattern of thought regarding how to turn "parts" of a concentration's expression into a measurable quantity.

   > **Example:** You need to make 100 mL of a 1:8 solution of alcohol-in-water.

**LESSON 31:** Concentration Expression: Identities and Dilution Factors

a. The solute/active ingredient in this solution is _____ and it makes up _____ parts.

b. The solvent/diluent in this solution is _____ and it makes up _____ parts.

c. How many total parts are in this solution?   Answer: _____

   i. How can you tell? What key words help you identify this? _____

d. How would you determine the amount of alcohol needed? _____

   i. Solve:

   Answer: _____

e. How would you determine the amount of water needed? _____

   i. Solve:

   Answer: _____

f. How can you check yourself? _____

---

**Example:** You need to make 200 mL of a 1:9 solution of insulin-to-water.

a. The solute/active ingredient in this solution is _____ and it makes up _____ parts.

b. The solvent/diluent in this solution is _____ and it makes up _____ parts.

c. How many total parts are in this solution?   Answer: _____

   i. How can you tell? What key words help you identify this? _____

d. How would you determine the amount of insulin needed? _____

   i. Solve:

   Answer: _____

e. How would you determine the amount of water needed? _____

   i. Solve:

   Answer: _____

**STOP AND PRACTICE:** Solve these for yourself:

1. A technician needs to make a 1:15 dilution of serum-in-water. The total volume must be 200 mL.

    a. What volume of serum is needed?

    Answer: _____

    b. What volume of diluent is needed?

    Answer: _____

2. A technician needs to make a 1:9 dilution of insulin-to-water. The total volume must be 500 mL.

    a. What volume of insulin is needed?

    Answer: _____

    b. What volume of water is needed?

    Answer: _____

## Understanding Dilution Factors

The term "dilution factor" is sometimes also referred to as the "solution dilution". It is the fraction that the solution was diluted by. Problems involving dilution factors can be easily solved using the ratio-proportion method.

> **Example:** Find the final concentration of a 60% solution that has been diluted by a factor of ¼

Step 1) Set the dilution factor as the known on the left side of the proportion.

$$\frac{1}{4}$$

Step 2) Set it equal to the diluted solution over the concentrated solution. Use cross multiplication to solve.

$$\frac{1}{4} = \frac{X\%}{60\%} \quad \text{where } X = \boxed{15\%}$$

## Stop and Practice:

1. Find the final concentration of a 20% solution that has been diluted by ¼.

    Answer: _____

## LESSON 31: Concentration Expression: Identities and Dilution Factors

2. Find the final concentration of a 50% solution that has been diluted by ½.

    Answer: _____

3. Find the final concentration of a 75% solution that has been diluted by 1/8.

    Answer: _____

4. Find the final concentration of a 25% solution that has been diluted 1/100.

    Answer: _____

Only a few slight changes are needed if both solution concentrations are given and the problem is asking for the dilution factor.

> **Example**: Find the dilution factor of a 20% solution that was originally 60%.

Step 1) Set the diluted solution over the concentrated solution as the known on the left side of the proportion.

$$\frac{20\%}{60\%}$$

Step 2) Set it equal to $1/X$. Use cross multiplication to solve, and report the answer as $1/X$.

$$\frac{20\%}{60\%} = \frac{1}{X} \text{ where } X = 4 \text{ so the answer is } \boxed{\frac{1}{4}}$$

## Stop and Practice:

1. Find the dilution factor of 10% solution that was originally 70%.

    Answer: _____

2. Find the dilution factor of a 20% solution that was originally 50%.

    Answer: _____

3. Find the dilution factor of a 10% solution that was originally 100%.

    Answer: _____

4. Find the dilution factor of a 5% solution that was originally 80%.

    Answer: _____

## Sampling the Certification Exam:

1. How many parts of water are in a 4:50 v/v insulin-in-water solution?
   a. 4
   b. 50
   c. 46
   d. 54

   Answer: _____

2. What is the dilution factor of a 25% solution that was originally 75%?
   a. 3
   b. 1/3
   c. ½
   d. ¼

   Answer: _____

3. What is the final concentration of a 90% solution that is diluted by 1/6?
   a. 5%
   b. 10%
   c. 15%
   d. 20%

   Answer: _____

4. How many parts of insulin are in a 1:5 v/v insulin-to-water solution?
   a. 1
   b. 4
   c. 5
   d. 6

   Answer: _____

# LESSON 31: Concentration Expression: Identities and Dilution Factors

5. A technician needs to make a 1:10 dilution of insulin-to-water. The total volume must be 80 mL. What volume of insulin is needed?

    a. 7.27 mL
    b. 8 mL
    c. 1 mL
    d. 10 mL

    Answer: _____

## Lesson 31 Content Check

1. If you dilute a 20% solution by ¼ what is the percent of the new diluted solution?

    Answer: _____

2. If a solution has been diluted from 75% to 25%, how much was it diluted?

    Answer: _____

3. Your pharmacy needs to make a 1:5 v/v solution of ammonia-in-water. If the total volume must be 120 mL, how many mL of ammonia do you need?

    Answer: _____

4. Your pharmacy needs to make a 1:10 v/v dilution of insulin-in-water, where the total volume must be 50 mL. How much insulin is needed?

    Answer: _____

5. Find the final concentration of a 50% solution diluted by ¼.

    Answer: _____

6. Your pharmacy needs to make a 1:10 v/v solution of insulin-in-water. If the total volume must be 500 mL, how many mL of water do you need?

    Answer: _____

7. If you dilute a 50% solution by ½, what is the percent strength of the new diluted solution?

    Answer: _____

8. **T / F**: A 1:20 v/v solution means that there are 1 parts of active liquid drug and 20 parts of diluent.

Answer: _____

9. If you have a 1:8 v/v solution of ammonia-to-water, how many parts are ammonia and how many parts are water?

Answer: _____

10. For each of the following v/v solutions, list the parts of active ingredient, inactive ingredient, and total:

| Solution: | Parts active ingredient | Parts inactive ingredient | Total parts |
|---|---|---|---|
| Bleach-to-water 1:10 | | | |
| Clorox-in-water 1:20 | | | |
| Epinephrine-to-water 1:5000 | | | |
| Alcohol-to-water 1:50 | | | |
| Insulin-in-water 1:10 | | | |

11. For each of the following solutions, list the amount (in mL) of the active ingredient and active ingredient:

a. 0.5 L of epinephrine-in-water, 1:5000 v/v

mL of active ingredient: _____
mL of inactive ingredient: _____

b. 300 mL of alcohol-in-water, 1:50 v/v

mL of active ingredient: _____
mL of inactive ingredient: _____

c. 10 oz of insulin-to-water, 1:10 v/v

mL of active ingredient: _____
mL of inactive ingredient: _____

d. 1 cup of urine-in-water, 1:100 v/v

mL of active ingredient: _____
mL of inactive ingredient: _____

# LESSON 31: Concentration Expression: Identities and Dilution Factors

e. ½ gal of serum-to-water, 1:40 v/v

        mL of active ingredient: _____

        mL of inactive ingredient: _____

f. 1 pt. of bleach-in-water, 1:4 v/v

        mL of active ingredient: _____

        mL of inactive ingredient: _____

# LESSON 32

# Alligations

Alligations is a method of solving various types of concentration and dilution problems. They are easily identified by the intent of the solution of the problem – ultimately, they require a pharmacy technician to make a diluted solution whose concentration lies somewhere in between solutions that are available, or in stock, in the pharmacy at the time. In order to best utilize this method, the concentration expression must be in the form of a percent, so if it is not, convert it before plugging it in.

To solve alligation problems, a tic-tac-toe board is used as the framework for determining the parts of the solutions to mix, and the total amount of each to measure out is solved using the ratio-proportion method. See the example below:

> **Example:** A pharmacy technician is instructed to make 1,000 mL of a 35% alcohol solution using a 20% alcohol stock solution and a 70% alcohol stock solution. How much of each stock solution is needed?

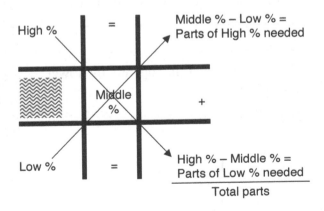

$$\frac{\text{parts of High \%}}{\text{total parts}} = \frac{X}{\text{total volume}}$$

Where $X$ = amount of high %

$$\frac{\text{parts of Low \%}}{\text{total parts}} = \frac{Y}{\text{total volume}}$$

Where $Y$ = amount of low %

# UNIT 5: Solution Calculations

## Description of Steps:

Step 1) Set up the tic-tac-toe board as seen above.

Step 2) Subtract the corners to find the parts of each.

Step 3) Add the parts together to find the total number of parts.

Step 4) Set up a ratio of parts over total parts for each strength. Alligation problems will usually have 2 answers each.

Step 5) Set each ratio equal to $X$ over the total volume to get the amount of each strength needed and solve using cross multiplication.

Step 6) Convert the amount to what is required in the problem if needed.

Step 7) Check that the amounts of each strength add up to the total volume needed.

Solution:

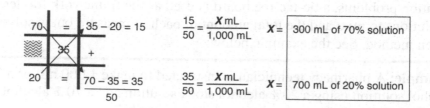

1. What did you notice about the identity (i.e., active ingredient; solute) of each solution in the previous problem? _____

   a. What conclusion can you draw about alligation problems due to this fact? _____
   _____
   _____

2. **T / F:** The higher percentage value must always go at the top, and the lower percentage value must always go at the bottom.

   a. Use the example from above to prove this below.

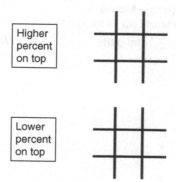

b. What HAS to stay consistent? _____

_____

3. **T / F:** You can only use the alligation method to solve problems involving volume.

4. Perform the following alligation problem:

   On the shelf in your pharmacy, you have hydrocortisone 10% and 1% cream. How much of each would you have to mix together to make 60 grams of a 2% cream?

   Answer: _____

5. Perform the following alligation problem:

   On the shelf in your pharmacy, you have a 1:100 dextrose solution and a 1:4 dextrose solution. How many milliliters of each would you have to mix together to make 4 L of a 1:10 dextrose solution?

   Answer: _____

   a. What 2 things are different in this problem vs. all the other ones before it?

   1 _____

   2 _____

6. If you have the following strengths of isopropyl alcohol solutions on your shelf – 5%, 35%, 50% and 70%. What are all the different combinations that you can mix together to make a 10% solution? (Hint: no math!)

   Combination 1: _____

Combination 2: _____

Combination 3: _____

a. Therefore, no matter what, you must always mix a _____ strength solution with a _____ strength solution to make the desired strength solution in an alligation.

## Understanding Solution Identities in Pharmacy Practice

1. What is $D_xW$? _____

   a. What is the percent of dextrose in each of the following:

      i. $D_{10}W$ = _____ % dextrose

      ii. $D_{7.5}W$ = _____ % dextrose

      iii. $D_{20}W$ = _____ % dextrose

      iv. $D_{50}W$ = _____ % dextrose

   b. Dextrose is a (**solid / liquid**) and therefore, a $D_xW$ solution will always be _____ / _____

   **STOP AND PRACTICE:** Solve the following for yourself

   You need to prepare 500 mL of a $D_{10}W$ solution using $D_5W$ and $D_{30}W$. How many milliliters of each are needed?

   Answer: _____

2. What is NS? _____

   a. What is the percent strength of NS always going to be? _____

   b. What is ½ NS? _____

      i. What is the percent strength of ½ NS?

   Answer: _____

c. What is ¼ NS? _____

   i. What is the percent strength of ¼ NS?

   Answer: _____

d. Saline, or salt, is a (**solid / liquid**) and therefore, a NS solution will always be _____ / _____

**STOP AND PRACTICE:** Solve the following for yourself

You need to prepare 25 mL of ½ NS. You only have ¼ NS and regular NS in stock. How many mL of each are required to make the required solution?

Answer: _____

3. What is SWFI? _____

   a. What is the percentage of dextrose in SWFI?   Answer: _____

   b. What is the percentage of alcohol in SWFI?   Answer: _____

   c. What is the percent strength of SWFI **always** going to be? Answer: _____

**STOP AND PRACTICE:** Solve the following for yourself

A prescriber orders 300 mL of a 30% alcohol solution. You are only able to find a 50% alcohol solution on your shelf, but you also have plenty of SWFI. How many mL of each are needed to complete the order?

Answer: _____

4. What is the percentage of petrolatum in hydrocortisone cream? _____
_____

   a. T / F: Many alligation problems that involve mixing solids will use petrolatum (0%) or aquaphor (0%) as a base, like those that involve mixing liquids will use water (0%) as a base.

**STOP AND PRACTICE:** Solve the following for yourself

Prepare 45 g of hydrocortisone 8% cream by using petrolatum and hydrocortisone 10% cream. How many grams of each are needed?

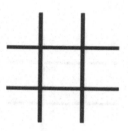

Answer: _____

## Sampling the Certification Exam:

1. To prepare 0.5 L of 12.5% dextrose solution from $D_{10}W$ and $D_{50}W$, how much of each is required?

   a. 45 mL 50% and 455 mL 10%

   b. 470 mL 50% and 30 mL 10%

   c. 360 mL 50% and 140 mL 10%

   d. 30 mL 50% and 470 mL 10%

Answer: _____

2. How many grams of a desoximetasone 1% cream will needed to be mixed with petrolatum to make 60 g of a desoximetasone 0.1% cream?

   a. 6 g of petrolatum, 54 g of desoximetasone 1%

   b. 7 g of petrolatum, 53 g of desoximetasone 1%

   c. 54 g of petrolatum, 6 g of desoximetasone 1%

   d. 53 g of petrolatum, 7 g of desoximetasone 1%

Answer: _____

3. Which of these solutions of sodium chloride (NaCl) in $H_2O$ could NOT be mixed together to make a 5% solution?

   a. SWFI & NaCl 20%

   b. NaCl 6% & NaCl 10%

   c. SWFI & NaCl 10%

   d. NaCl 2% & NaCl 8%

   Answer: _____

4. How many grams of dextrose are in 750 mL of $D_{10}W$?

   a. 750 g

   b. 75 g

   c. 10 g

   d. 7.5 g

   Answer: _____

5. How many grams of NaCl are in 500 mL of NS?

   a. 4.5 g

   b. 3.6 g

   c. 0.9 g

   d. 0.45 g

   Answer: _____

## Conclusions Regarding Multiple Choice Alligation Problems:

1. Look at the **WAY** alligation multiple choice question and answer choices are laid out in the above questions. What do you notice about the answer choices provided?
   _____

   a. Therefore, it is important that you must: _____
   _____

## Lesson 32 Content Check

1. Prepare a 400 mL solution of dextrose 8% in $H_2O$ using dextrose 5% in $H_2O$ and dextrose 50% in $H_2O$. How many mL of each are needed?

   Answer: _____

2. A prescription for 450g of Ammoniated Mercury Ointment 7% is received at your pharmacy. The pharmacy only has 3% and 10% Ammoniated Mercury available. Using alligation, calculate the amount of each available product needed to prepare this prescription.

Answer: _____

3. A nurse has a 70% alcohol solution and a 30% alcohol solution. How many mL of each should be used to make 400 mL of a 40% alcohol solution?

Answer: _____

4. Prescribers order: 50 mL of a 12% boric solution. The pharmacy carries only a 10% and an 80% boric solution. How much of each will you need to make the prescription?

Answer: _____

5. Prepare 120 g of a 2% hydrocortisone ointment using a 1% ointment and a 2.5% ointment. How much of each will you need to make the prescription?

Answer: _____

6. Prepare 90 g of triamcinolone 0.05% cream. In stock, you have 454 g each of triamcinolone 0.025% cream and triamcinolone 0.1% cream.

Answer: _____

7. Rx—Prepare 1 L of a 20% alcohol solution using the 90% alcohol and 10% alcohol that you have in stock. How many mL of each are needed?

Answer: _____

8. To prepare 500 mL of alcohol 20% using alcohol 70% and alcohol 10%, how many milliliters of each are needed?

Answer: _____

9. In order to prepare 50 mL of a 2.5% solution using a 10% solution and water, how many milliliters of each are needed?

Answer: _____

10. To prepare 2 kg of a 2% cream using a 30% cream and a 0.5% cream, how many g of each are needed?

Answer: _____

11. How many mL each of a 10% stock solution and SWFI are needed to make 2 L of a 0.5% solution?

Answer: _____

12. A prescriber has ordered 45 g of triamcinolone 0.075% ointment - a product that does not exist. Your pharmacist says that you can compound this drug by mixing together petrolatum and triamcinolone 0.1% ointment. How many grams of each will be needed to compound this order?

Answer: _____

# LESSON 33

# Solving Solution Problems

## Using the Alligation Method to Solve Dilution Problems

Let's use pictures to help us see how dilution REALLY works:

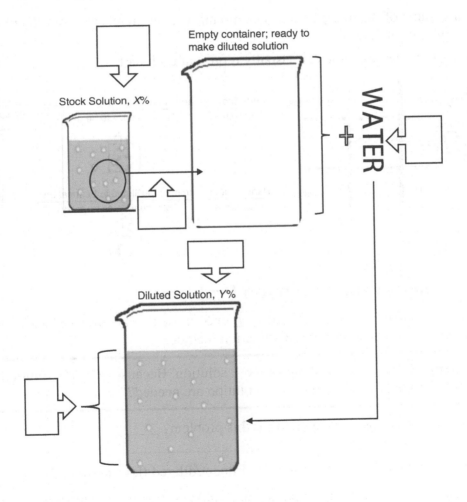

There are **5 parts** to any concentration/dilution problem. Label the pictures above with the part numbers described below:

1. <u>Stock solution strength</u>: more concentrated solution (higher percentage); what you physically **have** in the pharmacy to work with; Does not and cannot change
2. <u>Diluted solution strength</u>: less concentrated solution (lower percentage); often what is prescribed for or needs to be **made** in the pharmacy
3. <u>Volume of stock solution</u>: amount (usually in mL) that is physically taken out of the stock solution and placed into an empty beaker/container
4. <u>Volume of diluent (water)</u>: amount (usually in mL mL) of diluent, usually water, that is added to the volume of stock solution in the new beaker/container
5. <u>Total volume of the diluted solution</u>: amount (usually in mL mL) of diluent + amount (mL) of stock solution

Using the alligation method, you can visually represent each part:

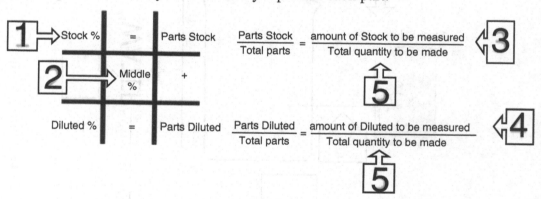

## Types of Concentration/Dilution Problems:

Type 1 – A problem that is asking for parts **3**, **4**, or **5** (those that are always looking for *volume*) can always be solved using the alligation method.

> **Example:** "You need 200 mL of a 6% solution. Because only 7% solution is in stock, how many mL of the stock solution are needed?"

a. Why/how can we use this method for this problem? _____
_____

i. What is the percent strength of the diluent/diluted (lower) strength? _____

Type 2 – A problem that is asking for parts **1** or **2** (always looking for *strength*) can always be solved using ratio/proportions (the dilution method).

> **Example:** If 16 mL of a 6% solution is diluted to 20 mL, what is the concentration of the diluted solution?

LESSON 33: Solving Solution Problems

## Solving Type 1 Concentration/Dilution Problems

First example – solving for part **3** (volume of stock solution)

> **Example:** You need 200 mL of a 6% solution. Because only 7% solution is in stock, how many mL of the stock solution are needed?

First - set up the problem using the tic-tac-toe board. Stock will always be mixed with water to make the diluted solution. Remember: water is always 0%! Then, solve using the alligation method steps you learned about earlier!

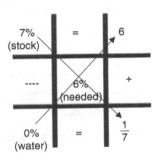

$$\frac{6}{7} = \frac{x \text{ mL}}{200 \text{ mL}} \quad x = 171.43 \text{ mL of } 7\%$$

*Tip: For this problem, there is no need to solve this part of the alligation as it only asked about the amount of stock

Second example – solving for part **4** (volume of diluent)

> **Example:** How many mL of diluent would it take to dilute a 25% stock solution down to 75 mL of a 5% solution?

Set up your tic-tac-toe board and plug in what you know, then solve.

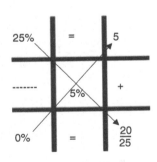

*Tip: For this problem, there is no need to solve this part of the alligation as it only asked about the amount of diluent

$$\frac{20}{25} = \frac{x \text{ mL}}{75 \text{ mL}} \quad x = 60 \text{ mL of diluent}$$

Third example – solving for part **5** (total volume of diluted solution)

> **Example:** You are using 100 mL of a 10% stock solution to make a 3% solution. How many mL of diluted solution can you make?

Set up your tic-tac-toe board and plug in what you know.

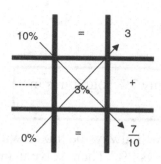

$$\frac{3}{10} = \frac{100 \text{ mL}}{x \text{ mL}} \quad x = 333.3 \text{ mL total volume}$$

Notice that the problem gave us the total of the stock solution used, and not the total volume, as it usually does. The key words here are "of", which lets you know that the 100 mL and the 10% go together.

"You are using **100 mL of a 10%** stock solution"

Additionally, you can tell that you are solving for the total volume because you will always be making the diluted solution. Therefore, anything referring to what needs to be "made" will always be referring to the total volume or total solution.

If you are having trouble "seeing" these types of problems, take a look at the following example and its rearrangement below.

**Straightforward alligation problem:**

> **Example:** How many mL of a 30% KCl solution and a 10% KCl solution are needed to make 2 L of a 12% solution?

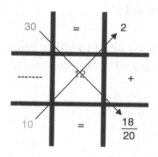

$$\frac{2}{20} = \frac{x \text{ L}}{2 \text{ L}} \quad x = 0.2 \text{ L or } 200 \text{ mL of } 30\% \text{ soln.}$$

$$\frac{18}{20} = \frac{x \text{ L}}{2 \text{ L}} \quad x = 1.8 \text{ L or } 1800 \text{ mL of } 10\% \text{ soln.}$$

**Rearranged – the SAME problem as above, asked in a different way:**

> **Example:** How many liters of a 12% solution can be made by mixing 200 mL of a 30% KCl solution with a 10% KCl solution?

# LESSON 33: Solving Solution Problems

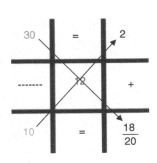

$$\frac{2}{20} = \frac{200 \text{ mL of } 30\% \text{ soln.}}{\text{Total Volume}}$$

where Total Volume = 2,000 mL, or 2 L

2nd step: "How many mL of the 10% KCl will be needed to achieve this total mixture?"

$$\frac{18}{20} = \frac{x \text{ mL}}{2,000 \text{ mL}} \quad x = 1,800 \text{ mL}$$

**STOP AND PRACTICE:** Solve the following:

1. How many mL of a 20% stock solution are needed to make 500 mL of a 5% solution?

Answer: _____

2. A 40% stock solution was diluted down to 10%. If the total volume of the mixture is 1,000 mL, how many mL of diluted solution were used?

Answer: _____

3. A pharmacy technician measures out 180 mL of $D_{50}W$ and mixes it with SWFI to make a $D_{20}W$ solution. What was the total quantity of the diluted solution?

Answer: _____

How can you easily identify these problems? What key words should you look for?
_____
_____

Therefore, you should always ask these questions when you know you are using the alligation method:

a. "Am I going to have one, or two answers?"

　　i. If only one answer is needed, a follow up question should be asked:

　　　　1. "Is the question asking for the volume of the stock solution used/needed, the volume of the diluted solution used/needed, or the total volume?

## Solving Type 2 Concentration/Dilution Problems

> **Example:** If 16 mL of a 6% solution is diluted to 20 mL, what is the concentration of the diluted solution?

Note that if you tried to set up an alligation, you'd get a funny picture like this:

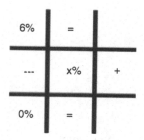

Instead, treat it like a ratio/proportion problem. Remember, we can compare anything and set it up as a ratio, and then set two equal ratios up as proportions if it makes _logical sense._

# LESSON 33: Solving Solution Problems

Therefore, the following <u>must</u> occur:

1. The units cancel each other out – either matching top to top and bottom to bottom, **OR** top to bottom and top to bottom

2. The logic of the comparisons you make is the same (i.e., part to whole equals part to whole).

*As a result, this is the logic:* $\dfrac{\text{Smaller volume (mL)}}{\text{Larger volume (mL)}} = \dfrac{\text{Smaller percent strength (\%)}}{\text{Larger percent strength (\%)}}$

Otherwise stated (but no need to memorize):

$$\boxed{\dfrac{\text{Volume of stock solution needed (mL)}}{\text{Volume of total diluted solution (mL)}} = \dfrac{\text{Diluted Solution \%}}{\text{Stock Solution \%}}}$$

Using this, plug in the information from above and solve:

$\dfrac{16 \text{ mL}}{20 \text{ mL}} = \dfrac{x\%}{6\%}$ where $x = \boxed{4.8\%}$

Want proof that this works? Let's use alligations to prove it:

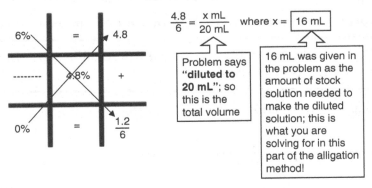

$\dfrac{4.8}{6} = \dfrac{x \text{ mL}}{20 \text{ mL}}$ where $x = \boxed{16 \text{ mL}}$

Problem says "diluted to 20 mL"; so this is the total volume

16 mL was given in the problem as the amount of stock solution needed to make the diluted solution; this is what you are solving for in this part of the alligation method!

**STOP AND PRACTICE:** Solve the following for yourself:

1. If 10 mL of a 15% stock solution is diluted to 20 mL, what is the concentration of the diluted solution?

    Answer: _____

2. If 30 mL of an unknown stock solution is diluted to 5% with a total volume of 200 mL, what must the concentration of the stock solution have been?

    Answer: _____

# UNIT 5: Solution Calculations

3. If 30 mL of a 60% stock solution is diluted to 90 mL, what is the concentration of the diluted solution?

   Answer: _____

4. If 150 mL of an unknown stock solution is diluted to 7% with a total volume of 260 mL, what must the concentration of the stock solution have been?

   Answer: _____

5. If 125 mL of a 80% stock solution is diluted to 200 mL, what is the concentration of the diluted solution?

   Answer: _____

How can you easily identify these problems? What key words should you look for? _____
_____

## Issues you might face with concentration problems

1. **When the problem uses a ratio strength instead of a percent strength**

   > **Example:** "To make 5,000 mL of a 1:2000 w/v solution, you would need how many mL of a 1:400 stock solution?"

   a. First thing – convert ratios to percent strengths

   $$1:2000 \rightarrow \% \quad \frac{1}{2000} = \frac{x}{100} \text{ where } x = 0.05\%$$

   $$1:400 \rightarrow \% \quad \frac{1}{400} = \frac{x}{100} \text{ where } x = 0.25\%$$

   b. Solve using alligation or ratio/proportion, depending on how the problem is worded

   i. In this case, alligations would work!

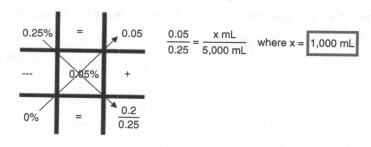

# LESSON 33: Solving Solution Problems    323

2. **When the problem doesn't look like a concentration problem at all**

   > **Example:** "Prepare one dose of 0.6 mg kanamycin in 2 mL diluted from a stock solution of 20 mg/mL. How much kanamycin and how much diluent are used?"

   a. First thing - look for key words.

   "Prepare one dose of 0.6 mg kanamycin in 2 (mL) diluted from (a stock solution) of 20 mg/mL. How much kanamycin and how much (diluent) are used?"

   b. Look for the strength of each solution (part *a* and *b* of any concentration problem).

   "Prepare one dose of (0.6 mg kanamycin in 2 mL) diluted from a stock solution of (20 mg/mL.) How much kanamycin and how much diluent are used?"

   c. Convert the strength of each solution to percent strength; it should start to look like a concentration problem now!

   i. Remember the definition of percent strength? Compare the strength of the solution to the definition of percent strength by using a proportion but watch your units!

   $$0.6 \text{ mg} \rightarrow \text{g} \quad \frac{0.006 \text{ g}}{2 \text{ mL}} = \frac{x}{100} \text{ where } x = 0.03\%$$

   $$20 \text{ mg} \rightarrow \text{g} \quad \frac{0.02 \text{ g}}{1 \text{ mL}} = \frac{x}{100} \text{ where } x = 2\%$$

   d. Look for the other parts of a typical concentration problem (parts *3*, *4*, and *5*) and determine what the question is asking for.

   "Prepare one dose of 0.6 mg kanamycin (in 2 mL) diluted from a stock solution of 20 mg/mL. How much kanamycin and how much diluent are used?"

   2 mL is the volume of the diluted solution. I know this because it says the 2 mL was **"diluted from** a stock solution". This is why we have to make sure we understand the concept!

   *The question is asking for the volume of stock used, and the volume of diluent used.*

   e. Solve using alligation or ratio/proportion, depending on how the problem is worded In this case, alligations would work!

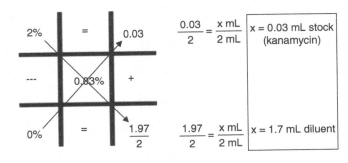

# UNIT 5: Solution Calculations

3. **When you don't know which is the stock solution and which is the diluted solution.**

   a. Always convert to a percent strength and then compare. The higher strength will always be the stock solution; the lower strength will always be the diluted solution

   b. Also, the smaller volume will always be the stock solution; the larger volume will always be for the diluted solution

      i. It only takes a *little* of the stock solution to make *a lot* of the diluted solution because we always add water!

4. **When you don't pay attention to the units.**

   a. This is the easiest thing to miss – using highlighters can help make sure you don't overlook anything!

5. **When extra information confuses you.**

   a. Always cross out information you don't need!

   > **Example:** "A patient needs 50 mL of 5% dextrose. You have 70% dextrose 1,000 mL on hand. How much concentrate will you need?"

      i. Upon first glance this might seem confusing. Read it again. What information do you NOT need? _____

         1. Why? _____

      ii. What is the question asking for? _____

## Sampling the Certification Exam:

1. A patient needs 175 mL of a 2% solution. Your stock solution is 10%. How much stock solution and diluent are needed?

   a. 75 mL stock, 100 mL diluent
   b. 45 mL stock, 130 mL diluent
   c. 35 mL stock, 140 mL diluent
   d. 140 mL stock, 35 mL diluent

   Answer: _____

# LESSON 33: Solving Solution Problems

2. To prepare 25 mL of a 10 mg/mL solution from a stock solution of 25 mg/mL, how much concentrate and how much diluent are needed?

   a. 15 mL stock, 10 mL diluent

   b. 10 mL stock, 15 mL diluent

   c. 20 mL stock, 15 mL diluent

   d. 20 mL stock, 5 mL diluent

   Answer: _____

3. If 50 mL of an unknown stock solution is diluted to 9% with a total volume of 400 mL, what must the concentration of the stock solution have been?

   a. 1.13%

   b. 72%

   c. 1.29%

   d. 63%

   Answer: _____

4. How many mL of a 1:7 w/v stock are needed to make 1.5 L of a 1:50 w/v solution?

   a. 210 mL of stock

   b. 1290 mL of stock

   c. 199 mL of stock

   d. 1301 mL of stock

   Answer: _____

5. 20 mL of a 1:50 w/v solution is diluted to 1000 mL. What is the percent strength of the new diluted solution?

   a. 0.02%

   b. 0.04%

   c. 0.4%

   d. 0.2%

   Answer: _____

# UNIT 5: Solution Calculations

## Lesson 33 Content Check

1. A prescriber ordered dextrose 8%, 400 mL. A pharmacy technician wants to make this solution from dextrose 20% and sterile water. How many milliliters of each solution are needed?

   Answer: _____

2. How much 23.4% sodium chloride should be added to 300 mL of sterile water to make a 3% sodium chloride solution?

   Answer: _____

3. A prescriber ordered 50 mL of a 5% dextrose solution. How many milliliters of 70% dextrose and how many milliliters of water does the technician need to make this solution?

   Answer: _____

4. A prescriber orders 60 g of lidocaine 4% $^W/_w$ ointment. How much of a lidocaine 10% $^W/_w$ ointment is required to compound his order? Note that you have plenty of aquaphor on the shelf to use.

   Answer: _____

5. A pharmacist needs 200 mL of a 6% NaCl solution. However, only a 7% solution of NaCl is available. How many mL of the 7% NaCl solution are needed to make 200 mL of the 6% solution?

   Answer: _____

6. You have been asked to make 90 g of a 3% hydroquinone cream using an 8% hydroquinone cream and aquaphor. How many mg of hydroquinone cream are needed?

   Answer: _____

7. How many mL of a 1:20 w/v stock solution are needed to make 1.5 L of a 1:500 w/v solution?

   Answer: _____

8. How many mL of a 1:40 w/v stock solution are needed to make 2 L of a 1:1000 w/v solution?

   Answer: _____

## LESSON 33: Solving Solution Problems

9. How many mL of a 1:20 w/v stock solution are needed to make 500 mL of a 1:75 w/v solution?

    Answer: _____

10. If a 40% solution of 50 mL was diluted to 200 mL, what is the percent of the diluted solution?

    Answer: _____

11. If there are 10 mL of an 8% solution, and water is added totaling to 50 mL, what is the new concentration of the diluted solution?

    Answer: _____

12. A 100 mL 50% w/v solution is diluted to 300 mL. What is the ratio strength of the new diluted solution?

    Answer: _____

13. A 25% w/v stock solution is used to make 100 mL of a 1:400 solution. How many mL of the stock solution were used?

    Answer: _____

14. A 50 mL 40% w/v solution is diluted to 500 mL. What is the ratio strength of the new diluted solution?

    Answer: _____

15. How many grams of benzocaine 2% ointment should you mix with 45 g of benzocaine 10% ointment to prepare a prescription of 4 oz of a 5% benzocaine w/w mixture?

    Answer: _____

16. Using a 15% w/w stock mixture of camphor in petrolatum, how many grams of the stock mixture will you need to make 150g of a 2% w/w mixture?

    Answer: _____

    a. In the question above, how many grams of petrolatum will be required?

    Answer: _____

17. A 1:10 stock solution is used to make 500 mL of a 1:2000 v/v solution. How many mL of the stock solution were used?

Answer: _____

a. How many mL of water are used in the above problem?

Answer: _____

18. What is the percent strength of epinephrine 1:5000, 1 mL diluted to a volume of 10 mL?

Answer: _____

19. How many milliliters of 4.2% sodium bicarbonate solution can be made from 100 mL of 8.4% concentration?

Answer: _____

20. The prescriber ordered 125 mL of a 2 mg/mL solution. Your stock solution is 15 mg/mL. How much stock solution and diluent are needed?

Answer: _____

# LESSON 34

# Math Literacy: Solution Problems

## What We've Learned

So far, only 3 types of problems have been discussed in this unit. Define them below:

| Type of Problem | How to identify | How to solve |
|---|---|---|
| Percent Strength | | |
| Alligation | | |
| Concentration / Dilution | | |

## Math Literacy

The goal of this exercise is to determine if you understand what a problem is asking, and whether you can set it up to solve. We will be doing the math to these problems at the end, but first, read the problem and then answer the questions that follow.

1. A pharmacy technician has a 200 mL solution containing 12 g of a medication. What is the percentage of the drug in this solution?

    *Percent Strength*
    *Alligation*
    *Dilution*

    a. What type of problem is this? How do you know? Circle/highlight key words.

    b. What type of mixture is this? How do you know? Circle/highlight key words.

    w/v  v/v  w/w

    c. Solve:

    Answer: _____

UNIT 5: Solution Calculations

2. **Ceftibuten is available in 10g / 100 mL concentration. How much diluent must be added to make one liter of a concentration of 50 mg / mL?**

   a. What type of problem is this? How do you know? Circle/highlight key words.

      i. Will there be one or two answers? How do you know? Circle/highlight key words.

         1. Which volume is it looking for? How do you know? Circle/highlight key words.

   b. Solve:

   | Percent Strength |
   | Alligation |
   | Dilution |

   | 1  2 |

   | Stock |
   | Diluent |
   | Total Volume |

   Answer: _____

3. **80 grams of hydrocortisone 3% ointment is added to a petroleum jelly base, making a total mixture weighing 1 pound. What is the new percent strength of hydrocortisone ointment?**

   a. What type of problem is this? How do you know? Circle/highlight key words.

      i. What additional step will this problem require? Why?

      ii. How do you know where to put the given percent strength in the proportion?

   b. Solve:

   | Percent Strength |
   | Alligation |
   | Dilution |

   Answer: _____

4. **What is the % W/V of a solution with a concentration of 60 mg / mL?**

   a. What type of problem is this? How do you know? Circle/highlight key words.

   b. What additional step will this problem require? Why?

   c. Solve:

   | Percent Strength |
   | Alligation |
   | Dilution |

   Answer: _____

LESSON 34: Math Literacy: Solution Problems   331

5. How much of 70% alcohol and how much of 20% alcohol are needed to make 500 mL of a 30% solution?

   *Percent Strength*
   *Alligation*
   *Dilution*

   1  2

   a. What type of problem is this? How do you know? Circle/highlight key words.

      i. Will there be one or two answers? How do you know? Circle/highlight key words.

   b. Solve:

   Answer: _____

6. A solution had an original concentration of 8%. It was then diluted and the final concentration was 2%. What is the dilution factor?

   *Percent Strength*
   *Alligation*
   *Dilution*

   a. What type of problem is this? How do you know? Circle/highlight key words.

   b. How is this problem slightly different than others like it?

   c. So, how will you solve this problem?

   d. How will your answer be reported?

   e. Solve:

   Answer: _____

7. A prescriber ordered potassium chloride 20%, 500 mL. The pharmacy stocks only 40% and 10% potassium chloride. How much of each is needed to make this solution?

   *Percent Strength*
   *Alligation*
   *Dilution*

   1  2

   a. What type of problem is this? How do you know? Circle/highlight key words.

      i. Will there be one or two answers? How do you know? Circle/highlight key words.

   b. Solve:

   Answer: _____

# UNIT 5: Solution Calculations

8. Cefoxitin is available in 15 g / 100 mL concentration. How much diluent, in L, must be added to make one liter of a concentration of 60 mg/mL?

   | Percent Strength |
   | Alligation |
   | Dilution |

   a. What type of problem is this? How do you know? Circle/highlight key words.

      i. Will there be one or two answers? How do you know? Circle/highlight key words.

         | 1   2 |

         1. Which volume is it looking for? How do you know? Circle/highlight key words.

            | Stock |
            | Diluent |
            | Total Volume |

         2. Solve:

   Answer: _____

9. How many grams of dextrose are needed to make 1500 mL of a 5% (w/v) solution?

   | Percent Strength |
   | Alligation |
   | Dilution |

   a. What type of problem is this? How do you know? Circle/highlight key words.

   b. Solve:

   Answer: _____

10. A pharmacy technician must prepare 4L of 28% iodine solution. The stock solutions available are 16% and 32%. How many milliliters of each solution are needed?

    | Percent Strength |
    | Alligation |
    | Dilution |

    a. What type of problem is this? How do you know? Circle/highlight key words.

       i. Will there be one or two answers? How do you know? Circle/highlight key words.

          | 1   2 |

    b. Solve:

    Answer: _____

# LESSON 34: Math Literacy: Solution Problems

11. A prescriber ordered dextrose 8%, 400 mL. A pharmacy technician wants to make this solution from dextrose 20% and sterile water. How many milliliters of each solution are needed?

   *Percent Strength*
   *Alligation*
   *Dilution*

   a. What type of problem is this? How do you know? Circle/highlight key words.

      i. Will there be one or two answers? How do you know? Circle/highlight key words.

      *1   2*

   b. Solve:

   Answer: _____

12. A pharmacy technician is compounding an IV solution containing dextrose with a concentration of 70% W/V. How many grams of dextrose are in 200 mL?

   *Percent Strength*
   *Alligation*
   *Dilution*

   a. What type of problem is this? How do you know? Circle/highlight key words.

   b. Solve:

   Answer: _____

13. How much 23.4% sodium chloride should be added to 300 mL of sterile water to make a 3% sodium chloride solution?

   *Percent Strength*
   *Alligation*
   *Dilution*

   a. What type of problem is this? How do you know? Circle/highlight key words.

      i. Will there be one or two answers? How do you know? Circle/highlight key words.

      *1   2*

         1. Which volume is it looking for? How do you know? Circle/highlight key words.

         *Stock*
         *Diluent*
         *Total Volume*

   b. Solve:

   Answer: _____

## UNIT 5: Solution Calculations

14. **A prescriber ordered 50 mL of a 5% dextrose solution. How many milliliters of 70% dextrose and how many milliliters of water does the technician need to make this solution?**

    a. What type of problem is this? How do you know? Circle/highlight key words.

    > Percent Strength
    > Alligation
    > Dilution

    i. Will there be one or two answers? How do you know? Circle/highlight key words.

    > 1   2

    b. Solve:

    Answer: _____

15. **How many grams of clindamycin 2% (w/w) ointment can be made from 600 g of clindamycin powder?**

    a. What type of problem is this? How do you know? Circle/highlight key words.

    > Percent Strength
    > Alligation
    > Dilution

    i. What type of mixture is this? How do you know? Circle/highlight key words.

    > w/v   v/v   w/w

    ii. How do you know where to put the given weight of drug in the proportion? Circle/highlight key words.

    b. Solve:

    Answer: _____

16. **A prescriber orders 60 g of lidocaine 4% W/W ointment. How much of a lidocaine 10% W/W ointment is required to compound his order?**

    a. What type of problem is this? How do you know? Circle/highlight key words.

    > Percent Strength
    > Alligation
    > Dilution

    i. Will there be one or two answers? How do you know? Circle/highlight key words.

    > 1   2

    1. Which volume is it looking for? How do you know? Circle/highlight key words.

    > Stock
    > Diluent
    > Total Volume

    b. Solve:

    Answer: _____

**LESSON 34:** Math Literacy: Solution Problems   **335**

17. What is the concentration of a 75 mL solution if it contains 12 mL of a 2.5% solution with water added to make the total 75 mL?

    > Percent Strength
    > Alligation
    > Dilution

    a. What type of problem is this? How do you know? Circle/highlight key words.

       i. How do you know where to put the given percent strength in the proportion?

    b. Solve:

    Answer: _____

18. A technician added an unknown volume of 5% NaCl solution into a beaker that already contained water. The total volume of the mixed solution in the beaker is 300 mL, and it's concentration is 4.5%. How many mL of the 5% NaCl solution were added into the graduate?

    > Percent Strength
    > Alligation
    > Dilution

    a. What type of problem is this? How do you know? Circle/highlight key words.

       i. Will there be one or two answers? How do you know? Circle/highlight key words.

    > 1   2

          1. Which volume is it looking for? How do you know? Circle/highlight key words.

    > Stock
    > Diluent
    > Total Volume

    b. Solve:

    Answer: _____

19. A pharmacist has 100 mL of a 2% solution. The pharmacist then dilutes it to 500 mL. What is the concentration of the 500 mL solution?

    > Percent Strength
    > Alligation
    > Dilution

    a. What type of problem is this? How do you know? Circle/highlight key words.

       i. How do you know where to put the given percent strength in the proportion?

    b. Solve:

    Answer: _____

20. A pharmacist needs 200 mL of a 6% NaCl solution. However, only a 7% solution of NaCl is available. How many mL of the 7% NaCl solution are needed to make 200 mL of the 6% solution?

   | *Percent Strength* |
   | *Alligation* |
   | *Dilution* |

   a. What type of problem is this? How do you know? Circle/highlight key words.

   | *1  2* |

       i. Will there be one or two answers? How do you know? Circle/highlight key words.

   | *Stock* |
   | *Diluent* |
   | *Total Volume* |

           1. Which volume is it looking for? How do you know? Circle/highlight key words.

   b. Solve:

   Answer: _____

21. A pharmacy technician adds water to make 160 mL of a 3.5% solution using 50 mL of a stock solution. What was the concentration of the stock solution used?

   | *Percent Strength* |
   | *Alligation* |
   | *Dilution* |

   a. What type of problem is this? How do you know? Circle/highlight key words.

       i. How do you know where to put the given percent strength in the proportion?

   b. Solve:

   Answer: _____

22. An IV containing dextrose has a concentration of 50% W/V. How many grams of dextrose are in 700 mL?

   | *Percent Strength* |
   | *Alligation* |
   | *Dilution* |

   a. What type of problem is this? How do you know? Circle/highlight key words.

   b. Solve:

   Answer: _____

**LESSON 34:** Math Literacy: Solution Problems    **337**

23. A pharmacist added an unknown volume of 9% KCl solution into a beaker that already contained water. If the total volume of the mixed solution in the graduate is 300 mL, and its concentration is 5%, how many mL of the 9% KCl solution were added into the graduate initially?

    | Percent Strength |
    | Alligation |
    | Dilution |

    a. What type of problem is this? How do you know? Circle/highlight key words.

        i. Will there be one or two answers? How do you know? Circle/highlight key words.

        | 1  2 |

            1. Which volume is it looking for? How do you know? Circle/highlight key words.

            | Stock |
            | Diluent |
            | Total Volume |

    b. Solve:

    Answer: _____

24. What is the percent strength of a guaifenesin (Robitussin) syrup made from 50 g of guaifenesin powder in 280 mL?

    | Percent Strength |
    | Alligation |
    | Dilution |

    a. What type of problem is this? How do you know? Circle/highlight key words.

    b. Solve:

    Answer: _____

25. Dilute 200 mL of chlorine in 500 mL of sterile water. What is the % v/v?

    | Percent Strength |
    | Alligation |
    | Dilution |

    a. What type of problem is this? How do you know? Circle/highlight key words.

        i. How do you know where to put the given amount of drug in the proportion? Circle/highlight key words.

    b. Solve:

    Answer: _____

# Unit 5 Content Review

1. How many milliliters of 2% lidocaine are needed to make a 300 mL solution containing 0.6 mg/mL?

    Answer: _____

2. How many milliliters of a 5% potassium solution are needed to make 120 mL of a 10 mg/mL concentration solution?

    Answer: _____

3. A 3:10 alcohol-in-water solution has _____ parts alcohol, _____ parts diluent, and _____ total parts.

4. 100 mL of a 5% sodium chloride solution is diluted to a volume of 1 liter. What is the new percent strength?

    Answer: _____

5. How many milliliters of 5% gentamicin solution can be made from 80 mL of 10% gentamicin?

    Answer: _____

6. How many milliliters of a 1:400 w/v stock solution are needed to make 4L of a 1:2000 w/v solution?

    Answer: _____

7. A laboratory technician needs to make a 1-in-10 dilution of serum-in-saline. The total volume must be 150 microliters. What volume of diluent is needed?

    Answer: _____

8. To make 3 liters of a 1:1000 w/v solution using a 1:500 w/v stock solution, how many milliliters of the stock solution are needed?

    Answer: _____

9. How many milliliters of 5% gentamicin solution can be made from 70 mL of 10% gentamicin?

    Answer: _____

10. How many grams of lidocaine 20% and 12.5% stock ointments must be combined to obtain 60 grams of a 15% lidocaine topical ointment?

Answer: _____

11. How many grams of Desoximetasone 0.1% cream can be made from 60g of a 1% concentration?

Answer: _____

12. Rx: zinc oxide 10% ointment 60 g. How many grams of zinc oxide 20% ointment and zinc oxide 5% ointment should you mix to prepare the order?

Answer: _____

13. How many milliliters of gentamicin 0.5% ophthalmic solution can be made from a 3 mL stock bottle of gentamicin in 2.5% ophthalmic solution?

Answer: _____

14. A 3:10 alcohol-to-water solution has _____ parts alcohol, _____ parts diluent, and _____ total parts.

15. Terconazole nitrate 8% cream 50 grams is diluted to 140 grams with a cream base. What is the new percent strength?

Answer: _____

16. Rx: coal tar 5% ointment 160 g. You have coal tar 10% ointment and coal tar 2% ointment. How many grams of each will you use to prepare the final product?

Answer: _____

17. How many grams of dexamethasone 2% ointment can be made from 160 g of dexamethasone 5% ointment?

Answer: _____

18. How many total grams of aminosyn are in a solution containing 300 mL of 7.5% and 250 mL of 5% concentrations?

Answer: _____

19. How much of 20% lidocaine and 30% lidocaine are needed to make 500 mL of a 28% concentration?

    Answer: _____

20. Dextrose 40%, 1000 mL is prescribed. The pharmacy only stocks 20% and 50% solutions. How much of each is needed to make the desired solution?

    Answer: _____

21. If you need 1500 mL of a 34% bleach solution, and the stock solutions available are 18% and 42%, how much of each will be needed?

    Answer: _____

22. A prescriber ordered an IV of 500 mL, 5% liquid solution. The pharmacy has a 20% solution. How many milliliters of sterile water are needed for this IV?

    Answer: _____

23. A prescriber ordered sodium chloride 500 mL, 0.45% liquid solution. The pharmacy technician makes this solution from 23.4% sodium chloride and sterile water. How many mL of each will be needed?

    Answer: _____

24. Hydrocortisone cream 3%, 60 g is ordered. The stock available is a 10% cream and a 1% cream. How many grams of each must be used?

    Answer: _____

25. A pharmacy technician wants to make 2 liters of 50% alcohol. The stock solutions are 1 liter of 20% alcohol and 7500 mL of 80% alcohol. How much of each alcohol is needed to make this solution?

    Answer: _____

26. A 1:20 insulin-in-water solution, 500 mL is requested from the prescriber. The technician should add _____ mL of insulin with _____ mL of water to make this prescription.

27. Dilute 200 mL of chlorine in 500 mL of sterile water. What is the % v/v?

    Answer: _____

28. How many mL of a 15% w/v solution can be made from 300 g of dextrose?

Answer: _____

29. A prescriber ordered 150 mL of a 4% w/v gentamicin solution. How much gentamicin powder is needed?

Answer: _____

30. A prescriber orders 40 g of lidocaine 2% w/w ointment. How much of a lidocaine 5% w/w stock ointment is needed to compound this order?

Answer: _____

31. The prescriber orders diphenhydramine 3% w/w cream. The directions ask for 60 mg per dose. How many grams of the cream is needed per dose?

Answer: _____

32. What is the final concentration if a saline solution consisting of 20% NaCl is diluted using a ¼ dilution?

Answer: _____

33. What is the final concentration if a saline solution containing 25% dextrose is diluted using a 1/10 dilution?

Answer: _____

34. What is the final concentration if a solution containing 50% dextrose is diluted using a 1/5 dilution?

Answer: _____

35. A pharmacist needs to make 400 mL of a 3% KCl solution by diluting a 6% KCl solution. How many mL of the 6% KCl solution are needed?

Answer: _____

36. A solution had an original concentration of 12%. It was then diluted. The final concentration was 4%. What is the dilution factor?

Answer: _____

37. A solution with a concentration of 8% is diluted ½ and then again by ¼. What is the final concentration?

Answer: _____

38. An alcohol solution is labeled as 20% v/v. How much alcohol is in 500 mL?

Answer: _____

39. A pharmacy technician needs to make 1,200 mL of a 7% NaCl solution by diluting a 12% solution of NaCl. How many mL of the 12% NaCl solution are needed?

Answer: _____

40. A pharmacy technician added a 4% NaCl solution into a graduate already containing a NaCl solution of unknown concentration. If the total volume of the mixed solution in the graduate is 200 mL and its concentration is 2.5% NaCl, how many mL of the 4% solution was added into the graduate?

Answer: _____

41. A pharmacist needs to dilute 60 mL of a 2.5% solution to 120 mL. What is the concentration of the 120 mL solution?

Answer: _____

42. A technician prepares 500 mL of 50% ethanol, using water to dilute a 90% ethanol solution. How many mL of water are needed?

Answer: _____

43. How many grams of a clindamycin 5% w/w ointment can be made from 300 mg of clindamycin powder?

Answer: _____

44. How many grams of talc should be used to prepare 400 g of a 5% w/w gel?

Answer: _____

45. How many grams of NaCl are needed to make 4000 mL of a 9% w/v solution?

Answer: _____

**LESSON 34:** Math Literacy: Solution Problems

46. What is the % v/v concentration of 60 mL phenol added to 400 mL sterile water?

    Answer: _____

47. What is the percent strength of a promethazine syrup made from 60 g of promethazine powder in 580 mL?

    Answer: _____

48. What is the % v/v concentration of 150 mL of alcohol if 750 mL of sterile water is added?

    Answer: _____

49. How many mL of a 12% w/v dextrose solution can be produced by using 30 g of dextrose?

    Answer: _____

50. If an 800 mL solution contains 150 g of Motrin, what is the percentage of Motrin in this solution?

    Answer: _____

# UNIT 6

# Hospital Pharmacy Math

# LESSON 35

# Powder Volume and Reconstitution

## Reconstitution

1. Why would we have to reconstitute a drug? _____
   _____

   a. The powder is _____ or "freeze dried" in order to _____
   _____

2. What does the strength of a liquid drug usually look like? Give an example with units.

3. What is a diluent? _____
   _____

   a. Give 3 examples of diluents often used to reconstitute powdered medications in pharmacy practice:

      1. _____
      2. _____
      3. _____

   b. What is another word for diluent? _____

4. Review the label below and answer the questions that follow:

   a. How can you tell that this is a drug that needs to be reconstituted? _____
   _____

   b. What diluent is needed for reconstitution? _____

   c. How much diluent is needed for reconstitution? _____

## Powder Volume:

Formula: | Powder Volume (PV) = Final Volume (FV) − Diluent (D) |

Use the principles of algebra to rearrange the equation putting the other two variables on the left side of the equal sign.

FV = _____

D = _____

1. How does this relate to v/v problems? _____
2. Powder Volume represents (**active / inactive**) drug.

Consider the following questions about powder volume problems:

1. **T / F** : Final volume is only affected by the amount of diluent added to a vial.

    a. As the diluent increases, the final volume (**increases / decreases**)

       i. Conversely, as the diluent decreases, the final volume (**increases / decreases**)

       ii. Explain:_____

2. **T / F** : Final concentration is only affected by the amount of diluent added to a vial.

    a. As the diluent increases, the final concentration (**increases / decreases**)

       i. Conversely, as the diluent decreases, the final concentration (**increases / decreases**)

       ii. Explain:_____

**LESSON 35:** Powder Volume and Reconstitution   349

3. **T / F** : Powder volume cannot change.

   a. Why/why not?_____
   _____
   _____

4. **T / F** : The actual drug powder that requires reconstitution has both a weight, and a volume.

   a. Describe:_____
   _____
   _____
   _____

   b. Therefore - **T / F** : The weight of the powder in a vial **and** the volume of the powder in a vial cannot change.

5. **T / F** : A drug's strength is not necessarily its total concentration.

   a. Why/why not?_____
   _____
   _____

   b. For example, review the following label:

   **METHOTREXATE INJECTION, USP**
   250 mg (25 mg/mL)
   CONTAINS PRESERVATIVE
   Cytotoxic Agent
   Sterile Isotonic Liquid
   NOT FOR INTRATHECAL USE
   10 mL Vial    Rx only

   i. What is the drug's strength? _____
   
   ii. Does that tell you the total amount of drug in the vial? _____
   
   iii. What is the total amount of drug in the vial? _____
   
   iv. What is the drug's total concentration? _____

v. Therefore, strength and concentration are related how? _____

_____

_____

## Solving All Powder Volume Problems

Powder volume problems will always be asking for either:

a. Powder volume

b. Final volume

c. Concentration

d. Diluent

It is important to remember a few things:

> 1. Powder weight and powder volume **<u>cannot</u>** change.
> 2. Concentration changes as you change the diluent.
> 3. Final volume changes as you change the diluent.

Solving problems asking for powder volume:

> **Example:** A 5 g vial requires that 8.6 mL of diluent be added to get a concentration of 250 mg/mL. What is the powder volume?

Step 1) Identification: How can you tell that this problem needs to be solved using the powder volume formula? _____

Step 2) First, always identify the parts of the problem:

Powder weight: _____

Concentration: _____

Final Volume: _____

Diluent: _____

Powder volume: _____

Step 3) Use the powder weight and the **concentration** to determine the **final volume** via the ratio-proportion method. Make sure the units match!

i. Why/how can you do this? _____

_____

## LESSON 35: Powder Volume and Reconstitution

Step 4) Plug in the **final volume** and the **diluent** into the powder volume equation to determine the **powder volume**.

Answer: _____

**STOP AND PRACTICE:** Try these on your own:

1. You add 8.7 mL to a 1.5 g vial to get a concentration of 125 mg/mL. What is the powder volume?

Answer: _____

2. Penicillin G potassium comes in a 1,000,000 unit vial. The medication needs to be reconstituted with 4.6 mL to have a concentration of 200,000 units/mL. What is the powder volume?

Answer: _____

3. You have a 15 g vial of medication with instructions to add 34.9 mL to have a concentration of 375 mg/mL. What is the powder volume?

Answer: _____

Solving problems asking for **final volume**:

> **Example:** An antibiotic for reconstitution says that in order to get a total concentration of 250 mg/mL, you should add 3.8 mL of bacteriostatic water to the 1.5 g of powder that has a powder volume of 2.2 mL. What is the **final volume**?

Step 1) Identification: How can you tell that this problem needs to be solved using the powder volume formula? _____

Step 2) First, always identify the parts of the problem:

    Powder weight: _____

    Concentration: _____

    Final Volume: _____

    Diluent: _____

    Powder volume: _____

Step 3) Notice that you have 2 of the 3 total parts of the powder volume equation (**powder volume** and **diluent**). Plug in and solve for **final volume**.

Answer: _____

**STOP AND PRACTICE:** Try these on your own:

1. An antibiotic has strength of 200 mg/tsp once reconstituted with 40 mL of diluent. If the powder volume is 5 mL, what will the final volume be?

Answer: _____

2. A vial of vancomycin 500 mg has a powder volume of 10 mL and instructions to add 90 mL of SWFI. What is the final volume, once reconstituted?

Answer: _____

3. A pharmacy technician is reading the instructions of a cefazolin label and measures out the 80 mL of diluent needed. If she adds this to the vial, containing 15 mL of powdered cefazolin, what will the vials total volume be?

Answer: _____

Solving problems asking for **concentration**: **NOTE: There are 2 types!!!!**

Type 1 – Simple

> **Example:** A 5 g vial requires 25 mL to be added. It has a powder volume of 5 mL. How many milligrams are in each milliliter of the final solution?

Step 1) Identification: How can you tell that this problem needs to be solved using the powder volume formula? _____

Step 2) First, always identify the parts of the problem:

   Powder weight: _____

   Concentration: _____

   Final Volume: _____

   Diluent: _____

   Powder volume: _____

Step 3) Use the **powder volume** and the **diluent** to determine the **final volume** via the powder volume equation.

Answer: _____

Step 4) Determine the **concentration** of the solution by comparing the total weight (powder weight) to the total volume (**final volume**) in a ratio. Reduce to lowest terms.

Answer: _____

i. Why can we do this? _____
_____
_____

Type 2 – Complex

> **Example:** A 10 g vial label says to add 20 mL of diluent to get 1 g per 2.5 mL. What **concentration** would you get if you added 35 mL?

Step 1) Identification: How can you tell that this problem needs to be solved using the powder volume formula? _____

Step 2) First, always identify the parts of the problem:

    Powder weight: _____

    Concentration: _____

    Final Volume: _____

    Diluent: _____

    Powder volume: _____

Step 3) Use the powder weight and the original **concentration** to determine the **final volume** via the ratio-proportion method. Make sure the units match!

Answer: _____

Step 4) Plug in the **final volume** and the **diluent** into the powder volume equation to determine the **powder volume**.

Answer: _____

Anticipate the follow up answer – remember, as diluent increases, concentration (**increases / decreases**)

Step 5) Now, it is important to remember that **powder volume** and powder weight cannot change, but the problem tells you that you now have a <u>new</u> diluent. Plug in the **powder volume** you just figured out, and your <u>new</u> **diluent** to get your new **final volume**.

Answer: _____

i. Why can we do this? _____
_____
_____

Step 6) Then, determine the **concentration** by putting the powder weight over (/) the **final volume**. Reduce as needed.

Answer: _____

Was the concentration affected the way you thought it would be? (**Y / N**) Therefore, you should always review your answers to see if they make logical sense!

**STOP AND PRACTICE:** Try these on your own:

1. A 10 g vial requires 88 mL of water to be added to reconstitute the antibiotic powder contained inside. It has a powder volume of 12 mL. How many grams are in each milliliter of the final solution?

    Answer: _____

2. A 250 mcg vial of phentolamine requires reconstitution with 0.9 mL of bacteriostatic water. If the powder volume of the phentolamine is 0.1 mL, what is the concentration of this reconstituted solution in mg/mL?

    Answer: _____

3. A 25 g vial label that the concentration of the drug is 750 mg/tbsp. if 120 mL of water is added. How would the concentration change if the pharmacy technician accidentally adds only 90 mL?

    Answer: _____

LESSON 35: Powder Volume and Reconstitution    355

4. A pharmacist asks a technician to dilute a reconstituted antibiotic that has a very strong flavor by adding 30 mL more water. If the concentration of the 75 mL bottle is 125 mg/5 mL, and the original instructions were to add 60 mL of water, what would the new concentration be?

Answer: _____

Solving problems asking for diluent: **NOTE: There are 2 types!!!!**

Type 1 – Simple

> **Example:** An oral medication requires reconstitution. The dose is 300 mg/tsp. The dry powder is 2.5 g with a volume of 9.6 mL. How much water do you add?

Step 1) Identification: How can you tell that this problem needs to be solved using the powder volume formula? _____

Step 2) First, always identify the parts of the problem:

    Powder weight: _____

    Concentration: _____

    Final Volume: _____

    Diluent: _____

    Powder volume: _____

Step 3) Use the powder weight and the **concentration** to determine the **final volume** via the ratio-proportion method. Make sure the units match!

Answer: _____

Step 4) Plug in the **final volume** and the **powder volume** into the powder volume equation to determine the **diluent.**

Answer: _____

Type 2 – Complex

> **Example:** A 25 g bulk vial label states that if you add 95 mL of a diluent, the concentration will be 1 g/5 mL. How much diluent would you add to get a concentration of 1 g/3 mL?

**NOTE** – this problem will utilize a similar approach to the "complex" type of concentration problem seen above**

Step 1) Identification: How can you tell that this problem needs to be solved using the powder volume formula? _____

Step 2) First, always identify the parts of the problem:

   Powder weight: _____

   Concentration: _____

   Final Volume: _____

   Diluent: _____

   Powder volume: _____

Step 3) Use the powder weight and the original **concentration** to determine the **final volume** via the ratio-proportion method. Make sure the units match!

Answer: _____

Step 4) Plug in the **final volume** and the **diluent** into the powder volume equation to determine the **powder volume**.

Answer: _____

Anticipate the follow up answer – remember, as concentration increases, diluent (**increases / decreases**)

Step 5) Now, it is important to remember that **powder volume** and powder weight cannot change, but the problem tells you that you now have a <u>new</u> **concentration**. Use the powder weight from above and the <u>new</u> **concentration** to determine the **final volume** via the ratio-proportion method. Make sure the units match!

Answer: _____

Step 6) Plug in the <u>new</u> **final volume** and the **powder volume** into the powder volume equation to determine the <u>new</u> **diluent**.

Answer: _____

Was the diluent effected the way you thought it would be? (**Y / N**)

## LESSON 35: Powder Volume and Reconstitution

**STOP AND PRACTICE:** Try these on your own:

1. An order arrives for 250 mg of Kefzol 100 mg/mL. Kefzol comes in a 500 mg vial and has a powder volume of 0.3 mL. How many milliliters of diluent must be added to reconstitute this medication?

   Answer: _____

2. A 2 g vial has a powder volume of 1.3 mL. You need to dilute the powder to a concentration of 150 mg/mL. How many milliliters of diluent must be added?

   Answer: _____

3. A 1 g vial of medication has a powder volume of 0.4 mL. How many milliliters of diluent need to be added to have a concentration of 50 mg/mL?

   Answer: _____

4. A 15 g vial label states that the concentration is 100 mg/mL if you add 107 mL of a diluent. How much diluent would you add to get a concentration of 150 mg/mL?

   Answer: _____

5. A 20 g bulk vial label states that if you add 77 mL of diluent the concentration will be 1 g/6 mL. How much diluent would you add to get a concentration of 1 g/5 mL?

   Answer: _____

## Sampling the Certification Exam:

1. Your bottle of Amoxil says to add 87 mL to the bottle to get a solution of 150 mg/tsp. The total amount in the bottle is 5 g. What is the powder volume?

   a. 46.3 mL
   b. 53.7 mL
   c. 79.7 mL
   d. 133.3 mL

   Answer: _____

2. You need to make an injectable solution with a final concentration of 350 mg/mL. You have a vial that contains 3.5 g with the instructions to add 6.5 mL. What is the powder volume?

   a. 3.5 mL

   b. 5.2 mL

   c. 6.5 mL

   d. 10 mL

   Answer: _____

3. You have added 4.2 mL of diluent to a 2g vial and now have a final volume of 10 mL. What is the powder volume?

   a. 2 mL

   b. 5 mL

   c. 4.2 mL

   d. 5.8 mL

   Answer: _____

4. The label of a 5g vial of medication says that if you add 5.5 mL of diluent, you will get a concentration of 500mg/5mL. What concentration per teaspoonful do you get if you add 7.5 mL?

   a. about 450mg/5ml

   b. about 480mg/5mL

   c. about 515mg/5mL

   d. about 508mg/5mL

   Answer: _____

5. An oral suspension once reconstituted is to have a dose of 250 mg/tsp, and the dry powder is 6 g with a powder volume of 21.8 mL. How much water must you add?

   a. 2.2 mL

   b. 98.2 mL

   c. 120 mL

   d. 128.2 mL

   Answer: _____

# Lesson 35 Content Check

1. What does "lyophilized" mean and what type of problems does it apply to?
2. What never changes in powder volume problems?

Answer: _____

3. The label of a 4 g vial states that you are to add 11.7 mL to get a concentration of 250 mg/mL. What is the powder volume?

Answer: _____

4. You are to reconstitute 1 g of dry powder. The label states that you are to add 9.3 mL of diluent to make a final solution of 100 mg/mL. What is the powder volume?

Answer: _____

5. You need a concentration of 375 mg/mL. Your vial contains 2 g with instructions to add 3.5 mL of diluent. What is the powder volume?

Answer: _____

6. A label states that a 5 g quantity of an antibiotic should be reconstituted with 8.7 mL saline for injection. The resulting concentration will be 500 mg/mL. What is the powder volume contained in the vial?

Answer: _____

7. An injectable medication comes packaged as a 1 g vial, and you want a final concentration of 125 mg/2 mL. The vial states that you are to add 14.4 mL diluent. What is the powder volume?

Answer: _____

8. You need to make an injectable solution with a final concentration of 375 mg/mL. You have a vial that contains 1.5 g with the instructions to add 3.3 mL of SWFI. What is the powder volume?

Answer: _____

9. A 10 g vial must have 45 mL of diluent added to it. It has a powder volume of 5 mL. How many mg will be in each milliliter of the final solution?

Answer: _____

10. The label of a 2 g vial states that you are to add 6.8 mL to get a concentration of 250 mg/mL. What is the powder volume?

    Answer: _____

11. In the above problem, how much diluent must you add to make a final concentration of 125 mg/mL?

    Answer: _____

12. A prescriber orders Amoxicillin 250 mg/5 mL suspension. When you pull the bottle off the shelf, you notice that the powder volume is 40 mL with directions to "Reconstitute with 60 mL of purified water". What will be the final volume of this mixture?

    Answer: _____

13. You have added 3.3 mL of diluent to a 1 g vial and now have a final volume of 4 mL. What is the powder volume?

    Answer: _____

14. In the above problem, how many mL of this medication will provide a 100 mg dose?

    Answer: _____

15. A 20 g bulk vial label states that if you add 106 mL of diluent, the concentration will be 1 g/6 mL. How much diluent would you add to get a concentration of 1 g/3 mL?

    Answer: _____

16. A 500 mg single-use vancomycin vial states "Reconstitute with 1.5 mL of sterile water for injection to obtain a concentration of 0.5g/2mL." The additional information on the back of the label states that the powder volume is 0.5 mL. What is the final volume of this vial once reconstituted?

    Aswer: _____

17. The label of a 6 g vial says that if you add 12.5 mL of diluent to the vial's contents, you will get a concentration of 1 g/2.5 mL. What concentration do you get if you add 2.5 mL?

    Answer: _____

**LESSON 35:** Powder Volume and Reconstitution — 361

18. A 3g vial of medication states that it has a powder volume of 15.8 mL. If the final concentration after reconstitution is 150mg/5 mL, how much diluent must be added?

    Answer: _____

19. You must reconstitute an antibiotic suspension before it can be dispensed to a patient. The dosage is 250 mg/tsp and the dry powder is 5 g with a powder volume of 8.6 mL. How much water must you add?

    Answer: _____

20. For an oral suspension you add 170 mL of water and get a final volume of 200 mL. If it contains 8 g of medication, how many milligrams will be in 1 tsp?

    Answer: _____

# LESSON 36

# IV Push and Continuous IV Infusions

Compare and contrast the following:

|  | Volume of drug administered | Amount of time of administration | Equipment Needed |
|---|---|---|---|
| **IV push (IV bolus)** | | | |
| **Continuous IV infusion** | | | |

1. What is the pharmacy technician's role in preparing IV push (IV bolus) medications?
   _____

   a. What does the final product look like (how is it given to the nurses)? _____

   b. The most common way to solve these problems is: _____

   _____

      i. What information do you have to have? _____

   > **Example**: A prescriber orders phenytoin 75 mg via IV push STAT. In the pharmacy you have a vial of phenytoin 50 mg/mL. How many mL will need to be drawn up to give to the patient?

   Answer: _____

2. What is the pharmacy technician's role in preparing IV infusion medications? _____

   _____

   a. What does the final product look like (how is it given to the nurses)? _____

   b. The most common way to solve these problems is: _____

   _____

      i. What information do you have to have? _____

   _____

## Intravenous Solution Bases

1. What are the 6 standard sizes of base solutions (i.e., how they will be stocked on your shelf)? _____
   _____

2. An IV bag with D$_\#$W means/will always have _____ _____ % in _____.

3. An IV bag with NS means/will always have _____ _____%
   a. Therefore, ½ NS is _____% and ¼ NS is _____%

4. What does LR, RL, or RLS stand for in terms of IV solutions? _____
   a. When/why is it needed? _____
   _____

5. T / F : An IV base can have a mixture of base solutions in it.
   a. Give an example: _____
      i. What does this mean? _____
      _____

**STOP AND PRACTICE:** Use your knowledge of percent solutions to solve the following:

1. D$_{7.5}$NS has ____ % dextrose and _____ % sodium chloride in 100 mL of solution.
   a. How much dextrose is in 250 mL of this solution?

   Answer: _____

   b. How much sodium chloride is in 250 mL of this solution?

   Answer: _____

2. D$_2$½ NS has ____ % dextrose and _____ % sodium chloride in 100 mL of solution
   a. How much dextrose is in 50 mL of this solution?

   Answer: _____

   b. How much sodium chloride is in 50 mL of this solution?

   Answer: _____

3. D$_5$W has ____ % dextrose and _____ % sodium chloride in 100 mL of solution
   a. How much dextrose is in 150 mL of this solution?

   Answer: _____

   b. How much sodium chloride is in 150 mL of this solution?

   Answer: _____

4. $D_{10}\frac{1}{4}$ NS has ____ % dextrose and _____ % sodium chloride in 100 mL of solution

   a. How much dextrose is in 500 mL of this solution?

   Answer: _____

   b. How much sodium chloride is in 500 mL of this solution?

   Answer: _____

5. LR has ____ % dextrose and _____ % sodium chloride in 100 mL of solution

   a. How much dextrose is in 1,000 mL of this solution?

   Answer: _____

   b. How much sodium chloride is in 50 mL of this solution?

   Answer: _____

## Intravenous Base Overfill

1. Why do manufacturers overfill intravenous bases? _____

   _____

   a. The loss of fluid is dependent on:

      i. _____

      ii. _____

      iii. _____

   b. The _____ the bag, the _____ the potential loss.

Examples of Overfill in IV bags:

| Size of Bag | Amount of overfill |
|---|---|
| 100 mL | 7 mL |
| 250 mL | 25 mL |
| 500 mL | 30 mL |
| 1,000 mL | 50 mL |

1. The amount of overfill can vary depending on the _____.

2. How can overfill affect a drug's concentration within an IV bag? _____

   _____

   a. What two areas of drug therapy should we be most concerned with overfill? _____

   _____

UNIT 6: Hospital Pharmacy Math

    i. Why? _____

  b. What type of sterile compounds are most affected by this phenomenon? _____

3. Fill in the blank for the general "rule" that most hospital policies implement regarding overfill:

   If the manufacturer's overfill plus _____ equals _____ or more over the stock label amount for the base IV solution, then the _____ should be removed before adding the drug volume.

   a. How do you determine if your bag requires removal of the base solution? _____

       i. How do you determine how much base solution to remove? _____

   b. Where can you find your institution's policy on overfill? _____

**STOP AND PRACTICE:** Determine whether the following IV orders will result in an overfill, and if so, how much base solution should be removed?

1. Drug order: Azithromycin 500 mg in 1,000 mL NS

   On hand: Azithromycin 500 mg vial for reconstitution; directions state to add 70 mL to reconstitute the vial

   a. How much overfill is typically seen in a 1,000 mL bag? _____
   b. How much drug will end up being added to the bag? _____
   c. How much total liquid would end up being added to the bag? _____
       i. Perform the calculations to determine if this is above the "10%" rule:

                                                                                                                         YES / NO

ii. So how much NS should be taken out of the bag?

Answer: _____

2. Drug order: Amiodarone 400 mg in 500 mL $D_{10}W$

   On hand: Amiodarone 400 mg vial for reconstitution; directions state to add 10 mL to reconstitute the vial

   a. How much overfill is typically seen in a 500 mL bag? _____
   b. How much drug will end up being added to the bag? _____
   c. How much total liquid would end up being added to the bag? _____
      i. Perform the calculations to determine if this is above the "10%" rule:

   **YES / NO**

   ii. So how much $D_{10}W$ should be taken out of the bag?

   Answer: _____

3. Drug order: Zyvox® 1.5 g in 100 mL $D_{2.5}W$

   On hand: Zyvox® 1 g/3 mL vial

   a. How much overfill is typically seen in a 100 mL bag? _____
   b. How much drug will end up being added to the bag? _____
   c. How much total liquid would end up being added to the bag? _____
      i. Perform the calculations to determine if this is above the "10%" rule:

   **YES / NO**

   ii. So how much $D_{2.5}W$ should be taken out of the bag?

   Answer: _____

# UNIT 6: Hospital Pharmacy Math

## Continuous Intravenous/Infusion Set Up

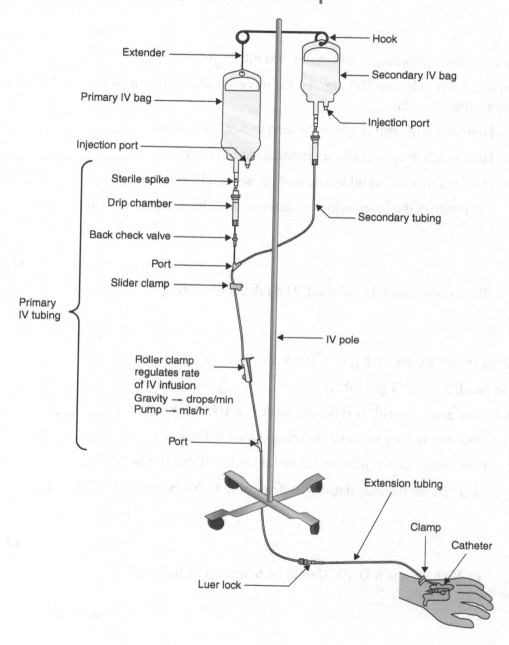

1. How can a nurse regulate the flow of the IV fluids? Define the following:
   a. Roller Clamp – _____
   _____
   b. Electronic Infusion Pump – _____
   _____

2. What is the difference between a primary IV bag and a piggyback? _____
   _____
   a. What does IVPB stand for? _____
   b. When is an IVPB used and why? _____
   _____
   _____

3. Concerning sizes:
   a. What does LVP stand for? _____
      i. The volume of an LVP is _____ mL or more.
      ii. LVP's are administered (**intermittently / continuously**)
   b. What does SVP stand for? _____
      i. The volume of an SVP is less than _____ mL.
      ii. SVP's are administered (**intermittently / continuously**)
   c. The label of the drug you are making will indicate a size/administration method so you can determine how to prepare the medication needed. No guess work!

4. What is a drop factor? _____
   a. Draw a picture that describes it.



   b. What are the units of drop factors always going to be? _____
   c. What are the standard macrodrop factors? _____
   d. What is the standard microdrop factor? _____
   e. Why does the size of drops change for different solutions? _____
   f. Why do we have to consider drop factor when determining a flow rate? ___
   _____

5. Finally, what is a flow rate in general terms?

   where the units could be: _____

   _____

   where the units could be: _____

   a. Sometimes a flow rate is called a(n) _____
   b. The most common unit of flow rates are _____ and _____.
   c. The best way to solve flow rate problems is by using _____

## Sampling the Certification Exam:

1. A microdrip IV drop set typically used for pediatric patients provides
   a. 15 gtt/mL.
   b. 10 gtt/mL.
   c. 20 gtt/mL.
   d. 60 gtt/mL.

   Answer: _____

2. Drugs that are administered by the direct IV injection route are often referred to as
   a. $D_5W$.
   b. hypodermics.
   c. IV push.
   d. saline locks.

   Answer: _____

3. How much dextrose is there if the order reads 1,000 mL $D_5W$?
   a. 5 g
   b. 50 g
   c. 500 g
   d. 5,000 g

   Answer: _____

4. Most hospital's P&P manual states that an IV bag for pediatric patients cannot have additives that exceed _____ of its base solution's volume due to the manufacturers overfill of all IV base solutions.

   a. 5%
   b. 7%
   c. 10%
   d. 15%

   Answer: _____

5. IV tubing with a drop factor of 15 means that the relationship is:

   a. 15 mL total
   b. 15 gtt/mL
   c. 15 gtt total
   d. 15 mL/gtt

   Answer: _____

## Lesson 36 Content Check

1. 0.5 L of $D_{10}$ ¼ NS has _____ g of dextrose and _____ g of sodium chloride.

2. How many mg of dextrose are in 50 mL of $D_7W$?

   Answer: _____

3. Ordered: 250 mL of $D_{10}W$. How many grams of sodium chloride are in this solution?

   Answer: _____

4. 300 mL of $D_{12}$ NS has _____ g of dextrose and _____ g of sodium chloride.

5. How many mg of sodium chloride are in 250 mL of a $D_2$ ¼ NS solution?

   Answer: _____

6. How much overfill is typically in a 250 mL IV base solution?

   Answer: _____

7. How much base solution should be removed in the following situation:

   Order: Zosyn® 3.375 g in 100 mL NS

   On Hand: Zosyn® 3.375 g/10 mL vial for reconstitution

   Answer: _____

8. How much base solution should be removed in the following situation:

   Order: Vancomycin 750 mg in 500 mL $D_5NS$

   On Hand: Vancomycin 250 mg/5 mL vial for reconstitution

   Answer: _____

9. 500 mL of $D_5$¼ NS is ordered. How much dextrose and sodium chloride, in grams, are in the bag?

   Answer: _____

10. 250 mL of NS is ordered. How many mg of sodium chloride is in the bag?

    Answer: _____

11. 500 mL of $D_5$0.33% NaCl is ordered. How many grams of dextrose and sodium chloride are in the bag?

    Answer: _____

12. 0.5 L of $D_{10}$¼ NS is ordered. How many grams of dextrose and sodium chloride are in the bag?

    Answer: _____

13. An example of an LVP is:

    Answer: _____

14. An example of an SVP is:

Answer: _____

Label the following pictures:

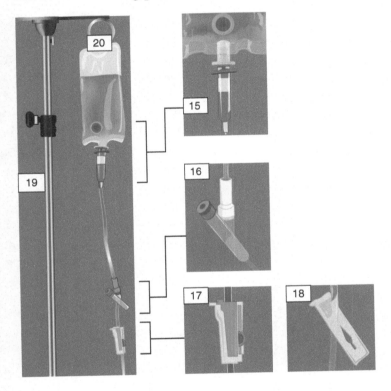

15. Answer: _____

16. Answer: _____

17. Answer: _____

18. Answer: _____

19. Answer: _____

20. Answer: _____

# LESSON 37

# Flow Rate Calculations

As discussed in the previous lesson, flow rate is, in general:

$$\frac{\text{Volume}}{\text{Time}}$$ where the units can be gtt, mL, or L
where the units can be s, min, hr

All flow rate problems can be easily identified by looking for the key words *"flow rate"* and should be solved with either the ratio-proportion method, or the dimensional analysis method. The thing all students should watch closely in all flow rate problems are the units! Pay attention to the units in the knowns and the units your answer needs to be in. Make sure at the end of the problem, you are solving for what is being asked.

## Basic Flow Rate Problems:

Some problems will separate the volume and the time aspect and simply ask for the rate per unit of time. The answer should be reported as a fraction.

> **Example:** A prescriber orders a 500 mL bag of NS to be run over 3 hours. What is the flow rate in terms of mL/hr?

1. Look at the units – what is the question asking for?

    a. Question asks for ***mL/hr*** meaning the answer needs to be reported as mL per 1 hour

2. Now you can put the volume over the time, and set equal to 1 hour:

    $\frac{500 \text{ mL}}{3 \text{ hr}} = \frac{X \text{ mL}}{1 \text{ hr}}$ where $X = 166.7$ mL so the answer would be $\boxed{166.7 \text{ mL/hr}}$

**STOP AND PRACTICE:** Try these:

1. A prescriber tells a nurse to hang a 1 L bag for a patient for a period of 8 hours. What would the flow rate be for this bag in mL/hr?

    Answer: _____

# UNIT 6: Hospital Pharmacy Math

2. A nurse needs to run a 250 mL IV bag for a pediatric patient over the period of 2 hours. What is the flow rate for this IV in mL/hr?

Answer: _____

3. A pediatric patient has been told that their 125 mL IV bag will take 1.5 hours to complete. What is the flow rate for this IV in mL/hr?

Answer: _____

Some problems will be very misleading by adding extra information that is unnecessary for the purposes of calculations.

> **Example:** A patient has an order for 500 mg of amoxicillin to be placed in a 100 mL bag and hung for a period of 2 hours. What is the flow rate in terms of mL/hr?

1. In identifying the type of problem, you should notice that the "500 mg of amoxicillin" is extra information because it has nothing to do with the flow (volume) rate (time) since mg is neither volume or time. <u>Cross it out so you don't feel compelled to use it!</u>

2. Look at the units – what is the question asking for?

   a. Question asks for **mL/hr** meaning the answer needs to be reported as mL per 1 hour

3. Now you can put the volume over the time, and set equal to 1 hour:

   $\frac{100 \text{ mL}}{2 \text{ hr}} = \frac{X \text{ mL}}{1 \text{ hr}}$ where $X = 50$ mL so the answer would be $\boxed{50 \text{ mL/hr}}$

**STOP AND PRACTICE:** Try these:

1. An order for a 1 g dose of Zosyn® arrives at the pharmacy with the instructions to add to a 100 mL IVPB to run for a total of ½ hr. What is the mL/hr that the pharmacy technician should instruct the nurse to set her electronic infusion pump to?

Answer: _____

2. A nurse enters an order to the pharmacy for 40 mEq of KCL in 500 mL NS to be infused over 6 hours. What is the flow rate, in mL/hr, for this IV bag?

Answer: _____

3. A prescriber orders 20,000,000 units of penicillin G in 1 L of $D_5NS$ to be infused over 8 hours. What is the flow rate, in mL/hr, for this IV bag?

Answer: _____

# LESSON 37: Flow Rate Calculations

Some problems give a unit of time within the problem that is different from the unit of time required in the solution (implying that you need to convert). In this case, dimensional analysis would be the best way to solve the problem.

> **Example:** An order for tetracycline 250 mg in a 50 mL IVPB is to be run over the course of 30 minutes. What flow rate in mL/hr should the technician tell the nurse to set the electronic infusion pump at?

1. Cross out extra information so you don't feel compelled to use it!
2. Look at the units – what is the question asking for?
    a. Question asks for **mL/hr** meaning the answer needs to be reported as mL per 1 hour **BUT** the problem gives the time component in terms of **minutes**
3. Now is where dimensional analysis becomes very useful:
    a. Set up initial flow rate as volume/time – most flow rate problems will start here (unlike other dimensional analysis problems discussed previously in the text that started with the unknown).

    $$\frac{50 \text{ mL}}{30 \text{ min}}$$

    b. Then, use dimensional analysis to convert FROM the unit of time the problem initially gives you, TO the unit of time the question asks for.

    $$\frac{50 \text{ mL}}{30 \text{ min}} \times \frac{60 \text{ min}}{1 \text{ hour}} = \boxed{\frac{100 \text{ mL}}{1 \text{ hr}}}$$

**STOP AND PRACTICE:** Try these:

1. Ordered: Ancef® 1 g in 100 mL $D_5W$ IVPB to be infused over 45 min. What is the flow rate in mL / hr?

    Answer: _____

2. Ordered: meropenem 1g in 100 mL $D_5W$ IVPB to be infused over 30 min by using an infusion pump. What is the flow rate in mL / hr?

    Answer: _____

3. Ordered: 1,500 mL Lactated Ringer's solution IV for 24 hr. How many milliliters per minute will the patient receive?

    Answer: _____

4. Ordered: 2,200 mL 0.45% NS per day by infusion pump. What is the flow rate in L / hr?

    Answer: _____

# Time and Supply Problems

Many problems will give you the volume of an IV bag, and a flow rate, then ask how long the bag will last. A simple proportion would help in this case.

> **Example:** A 1 L bag has 75 mL of medication added to it. The bag is running at 120 mL/hour. How long will it last?

1. Look at the units – what is the question asking for?

   a. Question asks for **mL/hr** meaning the answer needs to be reported as mL per 1 hour.

2. At this step, it is important to remember that volumes are additive, so when writing down your flow rate and determining time, be sure to incorporate this fact!

$$\frac{120\,mL}{1\,hr} = \frac{\overset{1075\,mL}{\cancel{(1\,L + 75\,mL)}}}{x\,hr} \text{ where } x = 8.96 \text{ hr rounded to } \boxed{9\,hr}$$

**STOP AND PRACTICE:** Find out how long the following bags would last:

1. 0.5 L flowing at 50 mL/hr

   Answer: _____

2. 200 mL with an infusion rate of 75 mL/hr

   Answer: _____

3. 100 mL with an infusion rate of 2 mL/min

   Answer: _____

4. 1 L flowing at 125 mL/hr

   Answer: _____

Some problems will give you a flow rate and a sig for administration, and then ask how many bags it would take to fill a cart for a particular nursing unit. Simple proportions, or just general knowledge of sig codes would help in this case as well.

> **Example:** An order comes in for a patient for cefepime, 325 mg in 100 mL NS over 30 min. TID. How many IVPB's of the drug should the technician prepare for a 24-hour cart fill?

TID = three times a day, so the technician would need to make 3,100 mL bags for that particular cart fill order.

**STOP AND PRACTICE:** Determine how many IV bags, and in what size, will be needed for the following 24-hour cart fill list for the Med-Surg floor of a hospital.

1. 1 L ½ NS q8h

    Answer: _____

2. 500 mL LR BID

    Answer: _____

3. 250 mL IVPB of Zosyn 1 g qam

    Answer: _____

4. 1,000 mL NS q6h – but the pharmacy is out of 1 L bags!

    Answer: _____

## Sampling the Certification Exam:

1. A 2 L bag is running at a rate of 120 mL/hour. How long will it last?

    a. 15 hours

    b. 17 hours

    c. 18 hours

    d. 20 hours

    Answer: _____

2. An IV bag is running at 50 mL/hour and only has 75 mL in the bag. How long will it last?

    a. 3 hours

    b. 2 hours

    c. 1.5 hours

    d. 0.5 hour

    Answer: _____

3. A patient is receiving 750 mg of vancomycin in 500 mL of NS over a period of 7 hours. What is the flow rate in milliliters per hour?

   a. 50 mL/hour
   b. 61 mL/hour
   c. 71 mL/hour
   d. 80 mL/hour

   Answer: _____

4. A volume of 75 mL is to be infused in 90 minutes using a mini-drip set. What is the flow rate in milliliters per hour?

   a. 50 mL/hour
   b. 57 mL/hour
   c. 72 mL/hour
   d. 100 mL/hour

   Answer: _____

5. If 1200 mL is to be infused at a rate of 100 mL/hour, how long will the bag last?

   a. 10 hours
   b. 11 hours
   c. 12 hours
   d. 13 hours

   Answer: _____

## Lesson 37 Content Check

1. Ordered: 1,500 mL Lactated Ringer's solution IV for 24 hr. How many milliliters per hour will the patient receive?

   Answer: _____

2. Ordered: 250 mL $D_5W$ over 3 hr by infusion pump. What is the flow rate in mL / hr?

   Answer: _____

3. Ordered: ceftriaxone 1 g in 50 mL NS IVPB to be infused over 20 min by using an infusion pump. What is the flow rate in mL / hr?

   Answer: _____

## LESSON 37: Flow Rate Calculations

4. Ordered: cefazolin 2 g IVPB diluted in 50 mL $D_5W$ to infuse in 15 min by an electronic infusion pump. What is the flow rate in mL / hr?

    Answer: _____

5. Ordered: 1800 mL Normal Saline IV to infuse in 15 hours by controller. What is the flow rate in mL / hr?

    Answer: _____

6. Ordered: 1,300 mL 0.45% NS per day. What is the flow rate in mL / min?

    Answer: _____

7. Ordered: 40 mEq KCl in 100 mL NS over 45 min. What is the flow rate in mL / hr?

    Answer: _____

8. Ordered: 850 mL $D_5W$ IV in 8 hr. What is the flow rate in mL / hr?

    Answer: _____

9. Ordered: 250 mL of 2 ½ % D/W IV. Infuse over a 24 h period. What is the flow rate in mL / hr?

    Answer: _____

10. Ordered: 650 mL $D_5W$ q8h IV. What is the flow rate in mL / hr?

    Answer: _____

11. Ordered: 850 mL 5% $D_5W$ IV in 8 hr. What is the flow rate in mL / hr?

    Answer: _____

12. Ordered: 1.5 L NS IV in 18 hours by controller. What is the flow rate in mL / hr?

    Answer: _____

13. Ordered: 250 mL of 2.5% D/W IV. Infuse over a 24 hr period. What is the flow rate in mL / hr?

    Answer: _____

14. You have an order for cefaclor 500 mg in 100 mL $D_5W$ to be infused over a 30-minute time period. What is the flow rate in mL/hr?

Answer: _____

15. You have an order for heparin 5,000 units in 50 mL NS to be infused over a 5-minute time frame. What is the flow rate (in mL/min)?

Answer: _____

16. You have an order for 8 mg of hydromorphone in a 50 mL IVBP to be administered over 15 minutes every 12 hours. What is the flow rate (in mL/min)?

Answer: _____

17. An order comes in for Augmentin® 500 mg to be run in a 50 mL bag over a 15-minute period. What is the flow rate in mL/hr?

Answer: _____

18. An order comes in for a 500 mL bag of levofloxacin 1500 mg to be run over a 3-hour period. What is the flow rate in mL/min?

Answer: _____

19. A 1000 mL bag has orders to be run evenly TID over a 24-hour period. What is the flow rate in mL/hr that should be set on each bag?

Answer: _____

20. A patient has an IV flowing at 25 mL/hr. If a 500 mL bag is hung, how long will this bag take to empty?

Answer: _____

# LESSON 38

# Drop Factor Calculations

All IV tubing is packaged and marked with the drop factor that it is calibrated to.

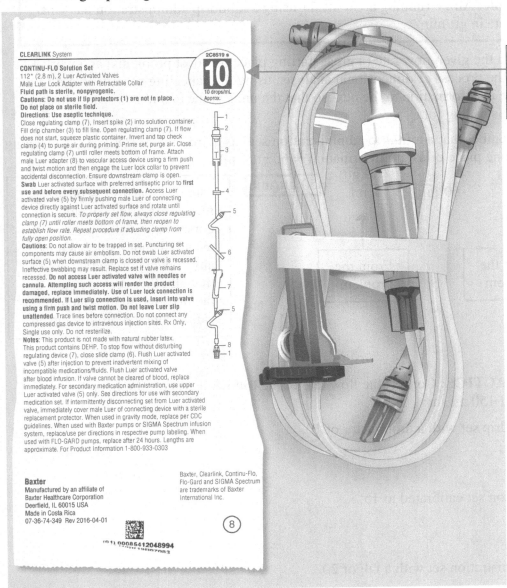

This package of IV tubing, sometimes called the "administration set" or "drop set", is calibrated to 10 gtt/mL as seen in the upper right-hand corner of the packaging.

Remember that drop factor (DF) is a value that relates the number of drops to a volume of 1 mL.

1. The units are always ____ / ____
2. What are standard macrodrip rates? _____
3. What is the standard microdrip (sometimes called "minidrip") rate? _____

As was the case for many of the flow rate problems in the previous lesson, drop factor problems can easily be solved using dimensional analysis.

> **Example:** An IV running at 50 mL/hr using a minidrip administration has what flow rate in gtt/min?

1. Start with the given flow rate (a great starting place for most problems) and work through the process of dimensional analysis, using the canceling of units to get from one piece of information to the next so that the units at the end match with what the question is asking for.

    a. Remember that a minidrip administration set has a drop factor (DF) of 60, meaning that one piece of useful information = 60 gtt/1 mL.

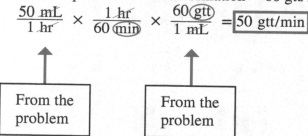

The fact that 1 hr = 60 minutes is often considered common knowledge, so usually is not mentioned in any conversion charts; note that all other conversion factors used in the calculation came from the problem itself.

**STOP AND PRACTICE:** Convert the flow rate of the following to gtt/min. Each has a prescribed rate of 150 mL/hr:

1. A macrodrip set with a drop factor of 10.

    Answer: _____

2. An infusion set calibrated to 15.

    Answer: _____

3. An administration set with a DF of 20.

    Answer: _____

4. A microdrip set.

Answer: _____

Often times, not all of the information given in the problem looks like it "traditionally" does in others.

- Flow rate may not be explicitly given, but any mention of **volume** and **time** can usually be put together to make one
- Drug concentration may not be explicitly given, by any mention of **strength/weight of drug** and **volume** can usually be put together to make one
  - Be sure to compare the same parts – part of drug strength to part of volume, or total drug amount to total volume

> **Example:** Ordered: 850 mL $D_5$ 0.45% NS q6h IV where the drop factor is 15 gtt / mL. Calculate the flow rate in drops per minute.

1. The original flow rate is not expressed explicitly, but it can be synthesized from the concept that flow rate = volume/time

    a. Therefore, the flow rate is = $\dfrac{850 \text{ mL}}{6 \text{ hr}}$

2. The rest of the problem should flow as most others did – using dimensional analysis. Generally, round to the nearest 1's place as it is not possible to get a partial drop.

$$\dfrac{850 \text{ mL}}{6 \text{ hr}} \times \dfrac{1 \text{ hr}}{60 \text{ min}} \times \dfrac{15 \text{ gtt}}{1 \text{ mL}} = \boxed{\dfrac{35 \text{ gtt}}{\text{min}}}$$

**STOP AND PRACTICE:** Try these on your own:

1. Ordered: Zosyn® 3 g in 100 mL $D_5W$ IV PB to be infused over 40 min.

    Drop factor: 10 gtt / mL

    What is the flow rate in gtt / min?

    Answer: _____

2. Ordered: famotidine 20 mg in 125 mL $D_5W$ IVPB to be infused over 45 min.

    Drop factor: 20 gtt / mL

    What is the flow rate in gtt / min?

    Answer: _____

3. Ordered: ciprofloxacin 500 mg in 200 mL D$_5$W IVPB to be infused over 30 min.

   Drop factor: 20 gtt / mL

   What is the flow rate in gtt / min?

   Answer: _____

4. Ordered: amoxicillin 2 g IVPB diluted in 100 mL D$_5$W to infuse in 30 min.

   Drop factor: 15 gtt / mL

   What is the flow rate in mL / hr?

   Answer: _____

Other types of flow rate problems just alter what they are asking for. Use dimensional analysis and watch the units carefully – the reported answer must have the units in the same order as they are asked in the question. Remember that relationships given within the problem cannot be separated, but they can be reoriented as needed/required by the logic of the process of dimensional analysis.

> **Example**: Procainamide hydrochloride at 60 mg/hr on an infusion device is ordered. Procainamide is available as an IV bag with a label of 1 g in 1,000 mL D$_5$W. What is the flow rate in mL / hr?

1. The drug's concentration, which is needed due to how the order is written, is not expressed explicitly, but it can be synthesized from the concept that concentration = weight/volume

   a. Therefore, the drug's concentration is = $\dfrac{1 \text{ g}}{1{,}000 \text{ mL}}$

2. The order is written in terms of milligrams but the drug's concentration is in grams – one of them must be converted. It does not matter which one as long as it is done correctly.

   a. Using the principles of KSMM, 1 g -> 1,000 mg, which now makes the drug's concentration = $\dfrac{1{,}000 \text{ mg}}{1{,}000 \text{ mL}}$

3. The question is asking for the mL/hr, indicating that the mL must be on top, so the drug's concentration must be flipped in order for the units to align the right way. This also is the only way that the only other piece of information given in the problem can be used, and the unit of "mg" can cancel out.

$$\dfrac{1{,}000 \text{ mL}}{1{,}000 \text{ mg}} \times \dfrac{60 \text{ mg}}{1 \text{ hr}} = \boxed{\dfrac{60 \text{ mL}}{1 \text{ hr}}}$$

Note that if the units were flipped another way, or other information was inserted, the logic behind the process of dimensional analysis would have yielded the wrong units in the answer. Therefore, it is important to let the units guide the problem, and not the other way around.

> **Example:** 1,000 mL 0.45% NS with heparin 25,000 units to infuse at 1000 units / hr was ordered. What is the flow rate in mL / hr?

1. The drug's concentration, which is needed due to how the order is written, is not expressed explicitly, but it can be synthesized from the concept that concentration = weight/volume

   a. Therefore, the drug's concentration is = $\dfrac{25,000 \text{ units}}{1,000 \text{ mL}}$

2. The question is asking for the mL/hr, indicating that the mL must be on top, so the drug's concentration must be flipped in order for the units to align the right way. This also is the only way that the only other piece of information given in the problem can be used, and the unit of "U" can cancel out.

$$\frac{1,000 \text{ mL}}{25,000 \text{ units}} \times \frac{1,000 \text{ units}}{1 \text{ hr}} = \boxed{\frac{40 \text{ mL}}{1 \text{ hr}}}$$

**STOP AND PRACTICE:** Try these:

1. Ordered: 500 mL $D_5W$ with heparin 25,000 U to infuse at 800 units / hr. What is the flow rate in mL / hr?

   Answer: _____

2. Ordered: 500 mL 0.45% IV with heparin 25,000 U to infuse at 500 units / hr. What is the flow rate in mL / hr?

   Answer: _____

3. Ordered: 500 mL $D_5W$ IV with heparin 40,000 U to infuse at 1,100 units / hr. What is the flow rate in mL / hr?

   Answer: _____

4. A patient is receiving 1 g of cefazolin in a 100 mL bag of $D_5W$ at a rate of 50 mL/hour. How many milligrams per hour will be administered?

   Answer: _____

The most complicated and detailed type of flow rate problems will only add in one additional factor: weight. When this comes up in a problem, start with that piece of information and use the logic of dimensional analysis to guide your calculations to the correct answer.

> **Example:** The prescriber orders 10 mg/kg/dose of a medication that is available as 25 mg/mL. The medication is to be mixed in a 100 mL bag of $D_5W$ and infused over 50 minutes using a mini-drip set. If the patient weighs 30 lb, what is the appropriate infusion rate in gtt/min?

1. The drug dose must be first be determined. Start with the weight, and use either information given in the problem or memorized conversion units to translate the information from unit to unit:

$$\frac{30 \text{ lb}}{1} \times \frac{1 \text{ kg}}{2.2 \text{ lb}} \times \frac{10 \text{ mg}}{1 \text{ kg}} \times \frac{1 \text{ mL}}{25 \text{ mg}} = 5.45 \text{ mL}$$

2. The problem says the drug is "mixed in a 100 mL bag", so the volume of the drug must be added to the volume of the bag in order to calculate the infusion rate in gtt/min. Additionally, the phrase "mini-drip set" alludes to a DF of 60 gtt/mL:

$$\frac{(5.45 \text{ mL} + 100 \text{ mL})}{50 \text{ min}} \times \frac{60 \text{ gtt}}{1 \text{ mL}} = \frac{126.54 \text{ gtt}}{1 \text{ min}} \text{ which rounds to } \boxed{\frac{127 \text{ gtt}}{\text{min}}}$$

**STOP AND PRACTICE:**

1. A medication is ordered for 15 mg/kg Q6 H given over 40 minutes using a 20-drop set. The patient weighs 75 lb and the medication is available as 40 mg/mL to be mixed in a 100 mL bag. What is the infusion rate in gtt/min?

    Answer: _____

2. A prescriber orders ketorolac 60 mg/day via IVPB to be divided and administered every 6 hours over a 20-minute infusion time frame.

    a. What dose of medication should be administered with each bag?

    Answer: _____

    b. If this medication is added to a 25 mL bag of $D_5W$, and the tubing has a DF of 15, how many gtt/min would the patient receive?

    Answer: _____

## LESSON 38: Drop Factor Calculations

## Sampling the Certification Exam:

1. A volume of 500 mL of whole blood is being infused into an anemic patient over 4 hours. The nurse is using a 10-drop set. What is the flow rate in drops per minute?

    a. 21 gtt/min

    b. 50 gtt/min

    c. 83 gtt/min

    d. 125 gtt/min

    Answer: _____

2. A patient is dehydrated and needs 1 L of lactated Ringer's solution infused over 6 hours. A 20-drop set is used. What is the flow rate in drops per minute?

    a. 56 gtt/min

    b. 50 gtt/min

    c. 17 gtt/min

    d. 5 gtt/min

    Answer: _____

3. A patient is receiving 20 mL over the next 30 minutes using a mini-drip set. What is the flow rate in drops per minute?

    a. 1 gtt/min

    b. 2 gtt/min

    c. 40 gtt/min

    d. 90 gtt/min

    Answer: _____

4. An order arrives for 1 L of NS with 20 mEq of KCl to be infused over 8 hours using a 20-drop set. What is the flow rate in drops per minute?

    a. 7 gtt/min

    b. 21 gtt/min

    c. 38 gtt/min

    d. 42 gtt/min

    Answer: _____

5. A patient is receiving 250 mg of Rocephin® in 100 mL NS over 1 hour and 30 minutes. A mini-drip set is used. What is the drop rate?

   a. 33 gtt/min

   b. 54 gtt/min

   c. 56 gtt/min

   d. 67 gtt/min

   Answer: _____

## Lesson 38 Content Check

1. If a 1500 mL bag is run every 5 hours with a drop factor of 20, what is the flow rate in gtt/min?

   Answer: _____

2. You have a 1,000 mL bag of $D_5W$ running at 125 mL/hr using a drop factor of 60. What is the flow rate in gtt/min?

   Answer: _____

3. The pharmacy has an order for an amlodipine IV 500 mL to run over a period of 8 hours, using a drop factor of 20. What should the pharmacy tell the nursing staff to run the IV flow rate at (in gtt/min)?

   Answer: _____

4. You have 500 mg of a drug on the shelf. The prescriber orders a 100 mL bag with a concentration of 50 mg/5 mL to be run over 3 hours using a drop factor of 10. What is the flow rate in gtt/min?

   Answer: _____

5. A prescriber orders 300 mL to be infused over 4 hours using a mini-drip set. What is the drop rate in gtt/hr?

   Answer: _____

6. If 40 mg of famotidine is mixed in 100 mL of $D_5W$ and infused over 30 minutes using a 10-drop set, what is the infusion rate in gtt/min?

   Answer: _____

**LESSON 38:** Drop Factor Calculations

7. What is the rate of infusion in gtt/min for a 100 mL bag with 25 mL of medication added that is infused over 25 minutes using a 10-drop set?

   (**HINT:** Be careful with this one! Remember, what do we know about volumes?)

   Answer: _____

8. If 1 g of medication is mixed into 50 mL of NS and is infused over 25 minutes using a 20-drop set, what is the infusion rate in gtt/min?

   Answer: _____

9. An IV that contains 2 mg in 150 mL is being infused at a rate of 25 mL/hour. How many micrograms will be administered per hour?

   Answer: _____

10. A patient is receiving 25 mEq of KCL in 1000 mL of NS over 18 hours using a 30-drop set. What is the flow rate in milliequivalent per minute?

    Answer: _____

11. An order arrives for 500 mL NS with 10 mEq KCl to be infused over 6 hours using a 10-drop set. How many milliequivalents per milliliter will be infused?

    Answer: _____

12. If 750 mg in 1000 mL is ordered to be infused over 6 hours using a mini-drip set, how many micrograms will be infused per drop?

    Answer: _____

13. If 12.5 mEq of electrolytes are added to 500 mL of NS and is infused over 12 hours using a 20-drop set, how many milliequivalents will be infused per drop?

    Answer: _____

14. A 1 L bag of $D_5W$ that contained 80 mEq of KCl has been discontinued after the patient has only received 650 mL.

    a. How many mEq of KCl has the patient received?

    Answer: _____

UNIT 6: Hospital Pharmacy Math

b. If 3 hours had passed before the order got discontinued, what was the flow rate set at?

Answer: _____

c. How many mEq per hour was the patient receiving?

Answer: _____

15. A prescriber orders ampicillin 2,000 mcg/kg IV STAT for a patient who weighs 165 lbs. The medication is to be added to a 50 mL IVPB of $D_5W$ to infuse for 20 minutes. The ampicillin comes as a 250 mg/5 mL vial.

   a. How much ampicillin should be added to the IVPB bag?

Answer: _____

   b. What is the flow rate in mL/hr that should be set?

Answer: _____

   c. How many mg of ampicillin is the patient receiving per minute?

Answer: _____

   d. If an administration set of 30 is used, what is the flow rate in gtt/min?

Answer: _____

16. A prescriber orders methylprednisolone 125 mg IV in a 100 mL $D_{10}W$ IV bag to run for an hour for a patient with asthma. The medication is supplied in a vial containing 250 mg/5 mL.

   a. How many mL of methylprednisolone would be added to the fluids to complete the order?

Answer: _____

   b. What is the flow rate in mL/min?

Answer: _____

   c. If the IV had a DF of 20, what would be the flow rate in gtt/min?

Answer: _____

LESSON 38: Drop Factor Calculations

17. A prescriber orders aminophylline as a loading dose of 5 mg/kg to be administered IVPB in 200 mL of fluid over 1 hour for a patient who weighs 154 lb. The available medication is a 10 mL vial containing 250 mg of aminophylline.

   a. How many mg of medication should the patient receive?

   Answer: _____

   b. How many mg of medication should the patient receive each hour if the order is changed and the dose is to be infused over 2 hours?

   Answer: _____

   c. If the infusion stops after 1.5 hours, how many mg of medication has the patient received?

   Answer: _____

18. A prescriber orders magnesium sulfate 10 g to be added to a 1 liter bag of LR. Magnesium sulfate is supplied to the pharmacy in a 50 mL vial labeled "50% Magnesium sulfate".

   a. What is the concentration of magnesium sulfate in mg/mL of the bag?

   Answer: _____

   b. What is the concentration of magnesium sulfate in mg/mL of the available solution?

   Answer: _____

   c. If this bag were to be infused over a 5 hour timeframe, how many milligrams of magnesium sulfate would the patient receive per hour?

   Answer: _____

19. A prescriber orders 1 L of $D_{10}$½NS to infuse over 10 hours.

   a. How many grams of dextrose is the patient receiving per hour?

   Answer: _____

   b. How many grams of sodium chloride is the patient receiving per hour?

   Answer: _____

20. A prescriber orders 1.5 L of LR solution to be infused over 8 hours. What is the flow rate in drops per minute if the administration set has a DF of 20?

Answer: _____

# LESSON 39

# Flow Rate Variations

1. What type of conditions would alter the set flow rate of an IV? _____
   _____

   a. How do nurses tell if a flow rate has been affected by these changes? _____
      _____

The formula to determine how much a flow rate has varied from its original intended rate is similar to the formula for percent error, as it is essentially calculating how much of a change has occurred. Therefore, we can use the following logic:

Formula: $$\% \text{ Variation} = \frac{|\text{New flow rate} - \text{Original flow rate}|}{\text{Original flow rate}} \times 100\%$$

Usually, this formula uses gtt/min as the standard flow rate for comparison. No matter what is used, the units of the new (recalculated) flow rate and the original flow rate must be the same when using this formula.

1. Most hospital policy and procedure manuals allow a nurse to adjust a patient's IV as long as the variation is _____ % or less.

   a. What occurs if the variation is more than that threshold? _____
      _____

   b. What occurs if there is no change in the patient's flow rate? _____
      _____

   c. Therefore, a nurse has only 3 courses of action:
      1. _____
      2. _____
      3. _____

# UNIT 6: Hospital Pharmacy Math

> **Example:** A 1 L bag of NS was ordered for a patient to run over the course of 4 hours using an administration set with a DF of 10. The nurse checks on the patient after 1 hour and sees that there are 850 mL left in the IV bag. What is the variation and action plan that the nurse should take?

1. Calculate the original flow rate in gtt/min:

   $$\frac{1{,}000 \text{ mL}}{4 \text{ hr}} \times \frac{1 \text{ hr}}{60 \text{ min}} \times \frac{10 \text{ gtt}}{1 \text{ mL}} = \frac{41.7 \text{ gtt}}{\text{min}}$$ which rounds to 42 gtt/min

2. Calculate the new flow rate in gtt/min:

   a. Remember that a flow rate is volume/time, and this type of problem is centered around the fact that the volume and the time changes from the beginning of the problem to the end. So, in order to calculate the new flow rate, determine how much volume and time remain based on the information given.

   Volume remaining: 850 mL – given in the problem directly

   Time remaining: 3 hours – determined by subtracting 1 hour from the total of 4 hours that the IV is supposed to run.

   Therefore, the new flow rate can be determined by the following calculation:

   $$\frac{850 \text{ mL}}{3 \text{ hr}} \times \frac{1 \text{ hr}}{60 \text{ min}} \times \frac{10 \text{ gtt}}{1 \text{ mL}} = \frac{47.2 \text{ gtt}}{\text{min}}$$ which rounds to 47 gtt/min

   > The DF cannot change because the same administration set is being used.

3. Use the formula to determine the percent variation:

   $$\frac{|47 - 42|}{42} \times 100\% = \boxed{11.9\%}$$

4. Based on the variation, determine the action plan:

   Since 11.9% is below 25%, the action plan should be that **the nurse adjusts the IV** to correct for the variation and ensure the IV completes on time as it was originally prescribed.

**STOP AND PRACTICE:** Try these:

1. **Ordered: 1200 mL $D_5W$ IV for 8 hours at 150 mL/hr using an administration set of 20**

   Original flow rate: _____ gtt/min

**After 3 hours, 400 mL of the IV bag has been infused.**

Solution remaining: _____ mL

Time remaining: _____ hr

Recalculated flow rate: _____ gtt/min

Variation: _____ %

Action: _____

2. Ordered: 900 mL of $D_5$ 1/2 NS IV for 12 hours at 75 mL/hr with a drop factor of 15

Original flow rate: _____ gtt/min

**After 7 hours, 100 mL of the IV bag remains.**

Solution remaining: _____ mL

Time remaining: _____ hr

Recalculated flow rate: _____ gtt/min

Variation: _____ %

Action: _____

3. Ordered: 1,500 mL of $D_5$ 1/2 NS for 10 hours at 150 mL/hr using a drop factor of 20.

Original flow rate: _____ gtt/min

**After 6 hours, 1,000 mL of the IV bag has been infused.**

Solution remaining: _____ mL

Time remaining: _____ hr

Recalculated flow rate: _____ gtt/min

Variation: _____ %

Action: _____

## Sampling the Certification Exam:

1. Hospitals usually consider a variation in flow rate for an IV of less than _____ as acceptable for the nurse to adjust.

    a. 10%

    b. 15%

    c. 25%

    d. 30%

    Answer: _____

2. An IV has begun to flow too fast, and the nurse calculates the variation at 30%. What is this nurse's course of action?

    a. Do nothing

    b. Adjust the IV flow rate

    c. Consult the prescriber

    d. Consult the pharmacy

    Answer: _____

3. A patient has been prescribed 2 L of hydration fluids to be run over the course of 12 hours using an administration set calibrated to 20 gtt/mL. A nurse checks on the patient after 6 hours of time has elapsed and notices that there are still 1.2 L remaining in the IV bag. What is the original flow rate of the IV in gtt/min?

    a. 55 gtt/min

    b. 56 gtt/min

    c. 66 gtt/min

    d. 67 gtt/min

    Answer: _____

LESSON 39: Flow Rate Variations    399

4. A patient has been prescribed 2 L of hydration fluids to be run over the course of 12 hours using an administration set calibrated to 20 gtt/mL. A nurse checks on the patient after 6 hours of time has elapsed and notices that there are still 1.2 L remaining in the IV bag. What is the new rate of the IV in gtt/min?

   a. 55 gtt/min
   b. 56 gtt/min
   c. 66 gtt/min
   d. 67 gtt/min

   Answer: _____

5. A patient has been prescribed 2 L of hydration fluids to be run over the course of 12 hours using an administration set calibrated to 20 gtt/mL. A nurse checks on the patient after 6 hours of time has elapsed and notices that there are still 1.2 L remaining in the IV bag. What is the degree of variation that has occurred?

   a. 16.4%
   b. 18.2%
   c. 19.6%
   d. 20.1%

   Answer: _____

## Lesson 39 Content Check

1. A patient is put on a slow IV drip of Amiodarone 1,500 mL for 15 hours using a drop factor of 20 to treat a heart condition.

   a. What is the flow rate in mL/hr?

   Answer: _____

   b. What is the flow rate in gtt/min?

   Answer: _____

   c. The nurse checks on the patient and notices that after 6 hours, 900 mL of the bag has been infused. She feels that the IV is running a little too fast, and decides to contact the pharmacy to determine her next step:

   i. What is the time remaining?

   Answer: _____

   ii. What is the volume remaining?

   Answer: _____

iii. What is the recalculated flow rate in gtt/min?

Answer:_____

iv. What is the % variation?

Answer:_____

v. What should the nurse to do?

Answer:_____

2. A patient in the ICU has a 1 L antibiotic IV to be infused for 3 hours using a drop factor of 15 to treat a severe infection.

   a. What is the flow rate in mL/hr?

   Answer:_____

   b. What is the flow rate in gtt/min?

   Answer:_____

   c. The nurse checks on the patient and notices that after 45 minutes, 300 mL of the IV bag has been infused:

      i. How much time is remaining?

      Answer:_____

      ii. How much volume is remaining?

      Answer:_____

      iii. What is the recalculated flow rate in gtt/min?

      Answer:_____

      iv. What is the % variation?

      Answer:_____

      v. What action should the nurse take?

      Answer:_____

3. Ordered: 2,000 mL of $D_{20}W$ for 15 hours at 133 mL/hr using a drop factor of 15.

   a. What is the flow rate in gtt/min?

   Answer:_____

   After 7 hours, 800 mL of the IV bag remains.

   b. How much time is remaining?

   Answer:_____

c. How much volume is remaining?

Answer:_____

d. What is the recalculated flow rate in gtt/min?

Answer:_____

e. What is the % variation?

Answer:_____

f. What action should be taken?

Answer:_____

4. Your order reads: 1000 mL NS for 9 hrs at 175 mL/hr with a drop factor of 15.

   a. What is the original flow rate in gtt/min?

   Answer: _____

   b. If after 2 hrs, the fluid remaining is 700 mL.

      i. What is the recalculated flow rate in mL/hr?

      Answer: _____

      ii. What is the recalculated flow rate in gtt/min?

      Answer: _____

   c. What is the variation?

   Answer: _____

   d. What is the action the nurse should take?

   Answer: _____

5. Your order reads: 300 mL fluids for 3 hours with a minidrip set.

   a. What is the original flow rate in gtt/min?

   Answer: _____

   b. If after 30 minutes, only 25 mL of the fluids have been infused.

      i. What is the recalculated flow rate in mL/hr?

      Answer: _____

      ii. What is the recalculated flow rate in gtt/min?

      Answer: _____

c. What is the variation?

Answer: _____

d. What is the action the nurse should take?

Answer: _____

## Unit 6 Content Review

### Short Answer

1. A prescriber orders penicillin G potassium 250,000 units q6h IM. The drug is available in a 1,000,000-unit multidose vial. The label reads to reconstitute the drug with sterile water.

   a. How much sterile water must be added to reconstitute 50,000 units/mL?

   Answer: _____

   b. How much water would you add to prepare 100,000 units/mL?

   Answer: _____

2. A pharmacy technician was directed to "add 7.8 mL of sterile water" to prepare a 10-mL multidose vial of a 100 mg/mL injection.

   a. How many milliliters of the reconstituted solution for injection would provide a 300 mg dose?

   Answer: _____

   b. What is the powder volume of the drug in this vial?

   Answer: _____

3. Refer to the cytarabine label below to answer the questions.

**CYTARABINE FOR INJECTION USP**

NDC 55390-133-01

LYOPHILIZED

See package insert for complete prescribing information.
Each vial contains cytarabine 1g and, if necessary, hydrochloric acid and/or sodium hydroxide for pH adjustment.
When reconstituted with 10 mL Bacteriostatic Water for Injection USP with benzyl alcohol, each mL contains 100 mg cytarabine. **Do not use a diluent containing benzyl alcohol for intrathecal and high dose investigational use.**
Store both powder and reconstituted solution at 25°C (77°F); excursions permitted to 15° to 30°C (59° to 86°F) [see USP Controlled Room Temperature].
Use reconstituted solution within 48 hours. Discard solution if a slight haze develops.

FOR INTRAVENOUS, SUBCUTANEOUS, OR INTRATHECAL USE

1 gram

Rx ONLY

Manufactured for:
Bedford Laboratories™
Bedford, OH 44146

Manufactured by:
Ben Venue Laboratories, Inc.
Bedford, OH 44146

CYB-VC05

    a. How much diluent is required for reconstitution?

       Answer: _____

    b. What type of diluent is to be used?

       Answer: _____

    c. How long can the reconstituted solution be retained?

       Answer: _____

4. You need to make a solution of Augmentin with a concentration of 500 mg/5 mL. The vial states that it contains 3 g of powder with the directions "Add 8 mL of diluent to reconstitute." What is the powder volume?

       Answer: _____

5. A patient has a prescription for Amoxicillin 125 mg/tsp. You know that you need to reconstitute the vial with 12.5 mL of sterile water, and you read that the vial has 1.5 g of powder inside. What is the powder volume?

       Answer: _____

6. The label of a 2 g vial states that to obtain a final concentration of 600 mg/tbsp, 35 mL of sterile water need to be added. What is the powder volume?

       Answer: _____

a. If you were to change the concentration of this vial to 500 mg/tbsp, how much diluent would you add?

Answer: _____

7. The label of a 6 g vial of medication states that if you add 15 mL of diluent, you will obtain a concentration of 1 g/10 mL. What concentration (in mg/mL) do you get if you add 25 mL?

Answer: _____

a. How many mL of this solution do you need to obtain a 125 mg dose?

Answer: _____

8. For a bottle of azithromycin, you add 25 mL to get a total volume of 40 mL. If it contains 0.75 g of medication, how many mg will be in 1 tsp?

Answer: _____

9. 300 mL of $D_{12}$ 0.9% NaCl is ordered. How many grams of dextrose and sodium chloride are in the bag?

Answer: _____

10. Ordered: 1.5 L NS IV in 18 hours by controller. What is the flow rate in mL / hr?

Answer: _____

11. Ordered: Augmentin® 1 g in 100 mL $D_5W$ IVPB to be infused over 45 min. What is the flow rate in mL/hr?

Answer: _____

12. Ordered: Zosyn® 3 g in 100 mL $D_5W$ IVPB to be infused over 40 min using an administration set with a drop factor of 10. What is the flow rate in gtt/min?

Answer: _____

13. Ordered: furosemide 120 mg IV push stat

On hand: furosemide 10 mg/mL

How many milliliters should the patient receive per dose?

Answer: _____

14. Ordered: levetiracetam 1g in 100 mL D₅W IVPB to be infused over 30 min by using an infusion pump. What is the flow rate in mL/hr?

Answer: _____

15. Ordered: carbamazepine 900 mg in 125 mL D₅W IVPB to be infused over 45 min. using an administration set with a drop factor of 20. What is the flow rate in gtt/min?

Answer: _____

16. Ordered: 100 mL IVPB ampicillin solution to infuse in 20 minutes using a drop factor of 40. What is the flow rate in gtt/min?

Answer: _____

17. Ordered: cephalexin 500 mg in 50 mL NS IVPB to be infused over 20 min by using an infusion pump. What is the flow rate in mL/hr?

Answer: _____

18. Ordered: ciprofloxacin 500 mg in 200 mL D₅W IVPB to be infused over 30 minutes with a drop factor of 20. What is the flow rate in gtt/min?

Answer: _____

19. Ordered: pantoprazole 40 mg IVPB diluted in 100 mL D₅W to infuse in 30 minutes using an administration set with a drop factor of 15. What is the flow rate in gtt/hr?

Answer: _____

20. Ordered: 1800 mL Normal Saline IV to infuse in 15 hours by controller. What is the flow rate in mL / hr?

Answer: _____

21. Ordered: 850 mL D₅0.45% NS q6h IV. The drop factor is 15. Calculate the flow rate in drops per minute.

Answer: _____

22. A prescriber orders 250 mL of D₅NS to be administered over 16 hours using a microdrip administration set.

   a. What is the flow rate in mL/hr?

Answer: _____

b. What is the flow rate in gtt/min?

Answer: _____

c. How many grams of dextrose are in the fluids?

Answer: _____

d. How many milligrams of sodium chloride are in the fluids?

Answer: _____

23. A prescriber orders conjugated estrogens 25 mg to be added to a 50 mL IVPB of $D_{2.5}W$ and run for 15 minutes. The drop factor for the administration set being used is 15.

   a. What is the flow rate in gtt/min?

   Answer: _____

   b. How many mcg of estrogen would be infused after 6 minutes of time had passed?

   Answer: _____

24. A prescriber orders lidocaine 300 mcg/kg IV for a heart patient who weighs 190 lb. The medication available is 4% lidocaine solution.

   a. How many mL of lidocaine should be added to the desired fluids?

   Answer: _____

   b. The prescriber wants this medication added to 20 mL of $D_5W$ for a period of 10 minutes. If the drop factor is 10, what flow rate (in gtt/min) nurse should set the infusion to?

   Answer: _____

   c. How many mg of lidocaine would the patient receive in 4 minutes?

   Answer: _____

25. A prescriber orders gentamicin 0.02 g to be added to 50 mL of NS set for an infusion time of 45 minutes. The drop factor is 20.

   a. What is the flow rate in gtt/min?

   Answer: _____

   b. How many mg of gentamicin would be added to the fluids?

   Answer: _____

   c. How many g of NaCl would the patient receive in 30 minutes?

   Answer: _____

   d. How many mcg of gentamicin would the patient receive in 15 minutes?

   Answer: _____

LESSON 39: Flow Rate Variations

26. A prescriber orders 1000 mL $D_5NS$ to be administered at 20 gtt/min. The DF is 15. What is the total infusion time for the ordered fluids?

Answer: _____

27. A prescriber orders amikacin 1 g in 100 mL $D_5W$ to infuse at 25 gtt/min using an infusion set calibrated to 60 gtt/mL.

   a. What is the total running time for the infusion rate in

   i. Minutes?

Answer: _____

   ii. Hours?

Answer: _____

   b. If the drug had a recommended infusion time of at least 1 hour, is this order safe for the patient?

Answer: _____

   c. How many mg of amikacin would the patient receive in 10 minutes?

Answer: _____

28. A prescriber orders oxacillin 12 g/day given in divided doses q3h IVPB. The medication is supplied in a 500 mg vial to be reconstituted to a final volume of 3 mL.

   a. How many mg of oxacillin would be given to the patient with each dose?

Answer: _____

   b. How many g of oxacillin would be given to the patient with each dose?

Answer: _____

   c. If this comes in an IV infusion prepared using what the pharmacy has in stock, how many vials of oxacillin would be necessary for one full day of treatment?

Answer: _____

   d. What would be the volume of medication to be added to a 100 mL bag of NS (assuming the 100 mL mini-bag contains the full dose)?

Answer: _____

   e. If the prescriber orders the medication to be given at 30 gtt/min, and the DF is 20, how long would this infusion take for completion in

   i. Hours?

Answer: _____

ii. Minutes?

Answer: _____

29. What is the conversion factor (fraction with appropriate units) for the following drop sets:

    a. Microdrip = _____

    b. administration set with a drop factor of 20 = _____

    c. macrodrip with a DF of 15 = _____

    d. minidrip set = _____

*Assume the flow rate is 200 mL / hr for the following questions:*

30. Administration set is 10 gtt / mL

    a. What is the drop factor?

    Answer: _____

    b. How many drops per minute will the patient receive?

    Answer: _____

31. Administration set is 15 gtt / mL

    a. What is the drop factor?

    Answer: _____

    b. How many drops per minute will the patient receive?

    Answer: _____

32. Administration set is 20 gtt / mL

    a. What is the drop factor?

    Answer: _____

    b. How many drops per minute will the patient receive?

    Answer: _____

33. Administration set is 60 gtt / mL

    a. What is the drop factor?

    Answer: _____

    b. How many drops per minute will the patient receive?

    Answer: _____

34. Your order reads: 3000 mL NS for 15 hrs at 200 mL/hr with a drop factor of 20.
    a. What is the original flow rate in gtt/min?

    Answer: _____

    b. If after 10 hrs, the fluid remaining is 500 mL.
        i. What is the recalculated flow rate in mL/hr?

        Answer: _____

        ii. What is the recalculated flow rate in gtt/min?

        Answer: _____

    c. What is the variation?

    Answer: _____

    d. What is the action the nurse should take?

    Answer: _____

# UNIT 7
# Pediatric Pharmacy Math

# LESSON 40

# Reference Text Pediatric Dose Calculations

Quick calculations of pediatric dosing are based on the child's weight (in kg) and then comparing it to dosing guidelines that are laid out in drug packaging inserts and references like the texts *Pediatric Dosage Handbook* or *Pediatric Injectable Drugs: The Teddy Bear Book* and online databases such as *Lexicomp, Clinical Pharmacology* or *IBM Micromedex NeoFax*. Therefore, it is important to remember the conversion from pounds to kilograms (1 kg = 2.2 lb). Another important conversion that has yet to be discussed but can aid your calculations is the conversion between pounds and ounces (1 lb = 16 oz) as many American health care systems use scales that measure partial pounds in terms of ounces. Neonates and infants require an exact weight to be known in order to provide a high degree of accuracy for their dosing requirements.

**STOP AND PRACTICE:** Calculate the following weights in kilograms:

1. 8,500 g

    Answer: _____

2. 12.3 lb

    Answer: _____

3. 7 lb, 8 oz.

    Answer: _____

4. 10 lb, 3 oz.

    Answer: _____

5. 22.3 lb

    Answer: _____

# UNIT 7: Pediatric Pharmacy Math

Most references will give either a _____ dosage or a _____ of doses.

1. What is the term for when a prescriber orders a dose below the minimum effective dose?
   _____

2. What is the term for when a prescriber orders a dose above the maximum dose allowable?
   _____

3. How can you determine if the dosage the prescriber ordered is safe? _____
   _____

Using the following example, many questions can be asked. All will utilize the ratio-proportion method to solve.

> **Example:** A 32 lb pediatric cancer patient requires a prescription for fentanyl. The recommended dosing according to a reference text is that pediatric patients can have 1-2 mcg/kg/dose to repeat q2-4h.

1. If the question, "What is the minimum effective dose for this child?" is asked, the solution would be:

   a. Find the weight of the child in kg.
   
   $$\frac{1 \text{ kg}}{2.2 \text{ lb}} = \frac{x \text{ kg}}{32 \text{ lb}} \quad \text{where } x = 14.54 \text{ kg}$$
   
   b. Calculate the minimum dose by using the ratio-proportion method.
   
   i. Since the range of doses is 1-2 mcg/kg/dose, the minimum dose would be 1 mcg/kg (the lower end of the range), giving:
   
   $$\frac{1 \text{ mcg}}{1 \text{ kg}} = \frac{x \text{ mcg}}{14.54 \text{ kg}} \quad \text{where } x = \boxed{14.54 \text{ mcg}}$$

2. If the question, "What is the maximum effective dose for this child?" is asked, the solution would be:

   a. Find the weight of the child in kg (as above).
   
   b. Calculate the maximum dose by using the ratio-proportion method.
   
   i. Since the range of doses is 1-2 mcg/kg/dose, the maximum dose would be 2 mcg/kg (the higher end of the range), giving:
   
   $$\frac{2 \text{ mcg}}{1 \text{ kg}} = \frac{x \text{ mcg}}{14.54 \text{ kg}} \quad \text{where } x = \boxed{29.01 \text{ mcg}}$$

3. If the question, "If the prescriber orders a 20 mcg dose, would this be safe for the patient?" is asked, the solution would be:

   a. Find the weight of the child in kg (as above).

   b. The calculations for the minimum and the maximum dose would need to be completed (as above) and then compared to the physician's dose to determine if it falls within the recommended dosing guidelines.

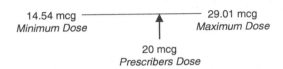

   The prescribed dose of 20 mcg falls in between the minimum and maximum dose, **so the ordered dose is safe.**

4. If the question, "If the drug is available in a 50 mcg/5 mL ampule, how many mL would give an ordered dose of 20 mcg?" is asked, the solution would be:

   a. This question is misleading! There are no unique or special calculations outside of a typical ratio-proportion question to be completed to find the solution. Never forget the basics!

   $$\frac{50 \text{ mcg}}{5 \text{ mL}} = \frac{20 \text{ mcg}}{x \text{ mL}} \quad \text{where } x = \boxed{2 \text{ mL}}$$

**STOP AND PRACTICE:** Calculate the following:

1. Ordered: Clindamycin 30 mg po TID for a child who weighs 34 lbs. The recommended pediatric dosage of clindamycin is 5-10 mg/kg/day in three divided doses.

   a. What is the child's weight in kg?

   Answer: _____

   b. What is the recommended minimum single dosage (mg/dose)?

   Answer: _____

   c. What is the recommended maximum single dosage (mg/dose)?

   Answer: _____

d. Write the range of dosages below. Draw an arrow to where the prescriber's ordered dosage lies.

e. Therefore, the dosage the prescriber ordered is (**subtherapeutic / appropriate / toxic**)

2. Ordered: Mirtazapine 1 mg po BID. Child weighs 55 lb. The recommended pediatric dosage of mirtazapine is 0.1-0.15 mg/kg/day in 2 equal doses.

   a. What is the child's weight in kg?

   Answer: _____

   b. What is the recommended minimum single dosage (mg/dose)?

   Answer: _____

   c. What is the recommended maximum single dosage (mg/dose)?

   Answer: _____

   d. Write the range of dosages below. Draw an arrow to where the prescriber's ordered dosage lies.

   e. Therefore, the dosage the prescriber ordered is (**subtherapeutic / appropriate / toxic**)

3. Ordered: Dilantin® 100 mg for a child who weighs 49 lbs. The recommended pediatric dosage of Dilantin® is 5-10 mg/kg/day.

   a. What is the child's weight in kg?

   Answer: _____

   b. What is the recommended minimum single dosage (mg/dose)?

   Answer: _____

   c. What is the recommended maximum single dosage (mg/dose)?

   Answer: _____

   d. Write the range of dosages below. Draw an arrow to where the prescriber's ordered dosage lies.

   e. Therefore, the dosage the prescriber ordered is (**subtherapeutic / appropriate / toxic**)

# LESSON 40: Reference Text Pediatric Dose Calculations

## Sampling the Certification Exam:

1. Convert 25 lb to kg
   a. 55 kg
   b. 12.6 kg
   c. 19 kg
   d. 11.4 kg

   Answer: _____

2. A neonate weighs 6 lb, 4 oz. How many kg does it weigh?
   a. 2.84 kg
   b. 2.91 kg
   c. 13.75 kg
   d. 14.08 kg

   Answer: _____

3. Pediatric dosing of acetaminophen is 10-15 mg/kg/dose for up to 5 doses per day. What is the minimum effective dose for a 22 lb child?
   a. 50 mg
   b. 75 mg
   c. 100 mg
   d. 150 mg

   Answer: _____

4. Pediatric dosing of ibuprofen has a maximum dosing threshold of 40 mg/kg/day. What is the maximum daily dose for an 8-month-old child weighing 16.2 lb?
   a. 145.5 mg
   b. 294.5 mg
   c. 704 mg
   d. 1,425.6 mg

   Answer: _____

5. A 16-month-old child experiencing an allergic reaction is admitted to the ER. The prescriber orders 12.5 mg of IV diphenhydramine. According to *Lexicomp*, diphenhydramine has a dosing range of 1-2 mg/kg/dose to repeat every 6 hours. If the child weighs 27.3 lb, the prescribed dose is:

    a. subtherapeutic

    b. appropriate

    c. toxic

    d. none of the above

    Answer: _____

## Lesson 40 Content Check

1. Convert 10 lb -> kg

    Answer: _____

2. Convert 5 lb, 2 oz -> kg

    Answer: _____

3. A 47 lb child with seizures is prescribed valproic acid. The reference material suggests a dosing guideline of 10-15 mg/kg/day in 2 divided doses.

    a. What is the minimum effective dose?

    Answer: _____

    b. What is the maximum effective dose?

    Answer: _____

    c. If a physician wrote for 250 mg BID, would this dose be safe?

    Answer: _____

    d. If the child was prescribed 200 mg BID and the drug was available in the pharmacy as a 250 mg/5 mL solution, how many mL would contain the appropriate dose?

    Answer: _____

**LESSON 40:** Reference Text Pediatric Dose Calculations

4. A 24.5 lb child with hypotension is prescribed phenylephrine. The reference material provides dosing guidelines that state IV bolus doses of 5-20 mcg/kg/dose every 10-15 minutes as needed to treat this condition.

   a. What is the minimum effective dose?

   Answer: _____

   b. What is the maximum effective dose?

   Answer: _____

   c. If the ER physician wrote an order for "0.2 mg IV q 10-15 min prn" would this dose be safe?

   Answer: _____

   d. If the child was prescribed a dose of 150 mcg every 10-15 minutes and the drug was available in the pharmacy as a 500 mcg/1 mL solution, how many mL would contain the appropriate dose?

   Answer: _____

5. Determine whether or not the following dosage is subtherapeutic, appropriate, or toxic according to the reference text recommendations:

   a. Ordered: clindamycin 300 mg IV q6h

   Patient weight: 46 lb

   Reference Text: 100 – 150 mg/kg/day in 4 divided doses

   Answer: _____

   b. Ordered: amiodarone 600 mg in 500 mL $D_5W$ over 24 hrs

   Patient weight: 72 lb

   Reference Text: 10 – 20 mg/kg/day

   Answer: _____

   c. Ordered: ibuprofen 50 mg po q4h prn pain

   Patient weight: 93 lb

   Reference Text: 10 – 40 mg/kg/day

   Answer: _____

   d. Ordered: lamotrigine 25 mg qd

   Patient weight: 62 lb

   Reference text: 0.15 – 0.3 mg/kg/day

   Answer: _____

# UNIT 7: Pediatric Pharmacy Math

6. Answer the following questions for a 51 lb child:

   Ordered: albuterol nebulizer

   Reference text: 0.15 mg/kg/dose every 20 minutes for 3 doses, then 0.15 – 0.3 mg/kg/dose every 1-4 hours; DNE 10 mg/dose

   a. How many milligrams should the patient get for the first 3 doses?

   Answer: _____

   b. What is the maximum amount (in milligrams) the patient can get per dose after the first three doses are given?

   Answer: _____

   c. How much would the patient have to weigh to exceed the single dosage threshold of 10 mg (assuming they are calculating based on the maximum use of the drug)?

   Answer: _____

7. Answer the following questions for a 23.4 lb child:

   Ordered: diazepam

   Reference text: 0.15 – 0.2 mg/kg/dose via slow IV; may repeat dose once in 5 minutes. Maximum dose of 10 mg/dose.

   a. How many milligrams should the patient get for the first dose?

   Answer: _____

   b. What is the maximum amount (in milligrams) the patient can get per dose?

   Answer: _____

   c. How much would the patient have to weigh to exceed the single dosage threshold of 10 mg (assuming they are calculating based on the maximum use of the drug)?

   Answer: _____

# LESSON 41

# Formula and BSA Pediatric Dose Calculations

## The Formula Methods:

All of the following rules equal to a pediatric dose (in theory):

| Clarks Rule | Youngs Rule | Fried's Rule |
|---|---|---|
| $\dfrac{\text{Weight (lb)}}{150 \text{ lb}} \times \text{Adult Dose}$ | $\dfrac{\text{Age (yrs)}}{\text{Age (yrs)} + 12} \times \text{Adult Dose}$ | $\dfrac{\text{Age (mo.)}}{150} \times \text{Adult Dose}$ |

1. Clarks Rule uses _____ whereas Young's Rule uses _____ in comparison to an adult dosage to determine the appropriate pediatric dose.

2. Who does Fried's Rule apply to? _____

3. Why do we no longer use Fried's, Clarks, or Young's rules to calculate pediatric dosing needs? _____
_____
_____
_____

## Body Surface Area (BSA):

1. What two methods are available for determining BSA? _____
_____

2. Why is BSA a more accurate way to determine appropriate dosage calculations for pediatric patients? _____
_____
_____
_____

# UNIT 7: Pediatric Pharmacy Math

| Age Group | Average BSA | |
|---|---|---|
| | Males | Females |
| Neonate (newborn) | 0.243 m² | 0.234 m² |
| 2 years | 0.563 m² | 0.540 m² |
| 5 years | 0.787 m² | 0.771 m² |
| 10 years | 1.236 m² | 1.245 m² |
| 13 years | 1.603 m² | 1.550 m² |
| 18 years | 1.980 m² | 1.726 m² |
| 20–79 years | 2.060 m² | 1.830 m² |
| 80+ years | 1.920 m² | 1.638 m² |

Figure 41.1 **Average BSA by age group and gender**

*Information sourced from "National Health and Nutrition Examination Survey". Cdc.gov. Retrieved 2019-11-5.*

Formula: $$\text{BSA (m}^2) = \sqrt{\frac{\text{weight (kg)} \times \text{height (cm)}}{3600}}$$

Note: for this formula, it is important to know the following conversions, just in case the information is given in measurements outside of the metric system:

1 kg = _____ lbs

1 inch = _____ cm

1 foot = _____ in

Note: Sometimes the " symbol is used to represent inches!

Note: Sometimes the ' symbol is used to represent feet!

Using the formula:

**Example:** Calculate the BSA for a child who is 18 lbs and 120 cm tall.

1. Make sure that the weight is in kg, and the height is in cm. If not, convert them:

$$\frac{1 \text{ kg}}{2.2 \text{ lb}} = \frac{x \text{ kg}}{18 \text{ lb}} \quad \text{where } x = 8.18 \text{ kg}$$

2. Plug each into the formula, and solve:

$$\sqrt{\frac{8.18 \text{ kg} \times 120 \text{ cm}}{3600}} = \boxed{0.52 \text{ m}^2}$$

LESSON 41: Formula and BSA Pediatric Dose Calculations    423

**STOP AND PRACTICE:** Calculate the BSA for the following:

1. A baby who is 60 cm tall and weighs 8 lbs has a BSA of:

   Answer: _____

2. A toddler who is 71 cm tall and weighs 34 lbs has a BSA of:

   Answer: _____

3. A toddler who is 33 inches tall and weighs 18 kg has a BSA of:

   Answer: _____

1. What is a "West" nomogram? _____
   _____
   _____

   a. What information is needed for this chart? _____

   b. How do you use a nomogram for a child who:

      i. Is of normal height for their weight. _____
         _____
         _____

         1. Highlight this area in figure 42.2.

      ii. Is underweight/overweight. _____
          _____
          _____

## BSA Dosing in Reference Texts

Reference texts and product inserts are starting to include dosing guidelines by BSA due to its popularity in calculating dosages. The references are given in terms of milligrams, micrograms, or other units per $m^2$. The ratio-proportion method is a simple approach to solving all of these problems once the BSA is calculated.

> **Example:** A 3-year-old patient with leukemia is prescribed vincristine. The recommended dosing guidelines for this drug is 2 mg/$m^2$. If this patient is 32 lb and 92 cm tall, what is their recommended dose in mg?

# UNIT 7: Pediatric Pharmacy Math

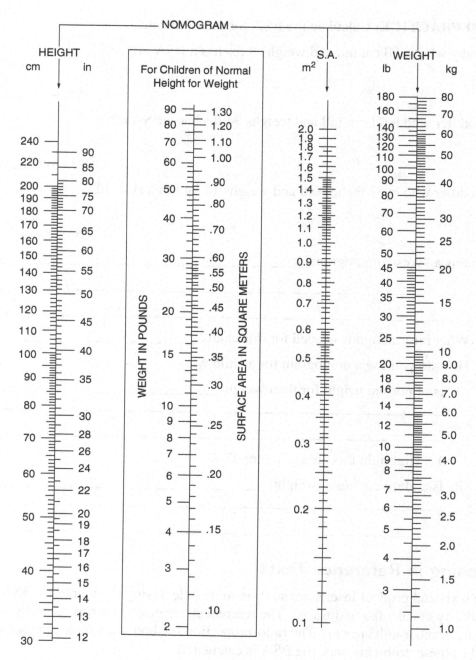

Figure 42.2  **A West Nomogram**

*Information sourced from: https://www.austincc.edu/rxsucces/ped3.html. Retrieved 2021-24-8.*

# LESSON 41: Formula and BSA Pediatric Dose Calculations

1. Calculate their BSA.
   a. In order to do this, the weight must be converted to kg.
   
   $$\frac{1 \text{ kg}}{2.2 \text{ lb}} = \frac{x \text{ kg}}{32 \text{ lb}} \quad \text{where } x = 14.54 \text{ kg}$$
   
   b. Then, plug their weight and height into the BSA formula:
   
   $$\sqrt{\frac{14.54 \text{ kg} \times 92 \text{ cm}}{3600}} = 0.61 \text{ m}^2$$

2. Set up a proportion of the recommended BSA dosing equal to their BSA to determine the appropriate dose.

   $$\frac{2 \text{ mg}}{1 \text{ m}^2} = \frac{x \text{ mg}}{0.61 \text{ m}^2} \quad \text{where } x = \boxed{1.22 \text{ mg}}$$

**STOP AND PRACTICE:** Calculate the following:

1. A 41 lb pediatric patient is ordered to take a medication with a reference dose of 10 mg/m². If this patient is 113 cm tall, what is their dose of medication in mg?

   Answer: _____

2. An infant with a BSA of 0.25 m² is prescribed a medication with a reference dose of 0.5 mg/m². What is their dose of medication in mcg?

   Answer: _____

3. A 46 in tall child who weighs 80 lb is prescribed a medication that has a reference dose of 23.5 mg/m². What is their dose of medication in g?

   Answer: _____

# UNIT 7: Pediatric Pharmacy Math

## Sampling the Certification Exam:

1. The BSA of a child can be quickly determined by using:
   a. A West Nomogram
   b. A Pediatric Reference Text
   c. *Lexicomp*
   d. Clark's Rule

   Answer: _____

2. The units of BSA are usually:
   a. $m^3$
   b. $mcg^2$
   c. $mg^2$
   d. $m^2$

   Answer: _____

3. 1 inch is equal to
   a. 1 cm
   b. 1.5 cm
   c. 2.54 cm
   d. 3 cm

   Answer: _____

4. What is the BSA of a 40 lb child who is 90 inches tall?
   a. 0.67 $m^2$
   b. 1.00 $m^2$
   c. 1.07 $m^2$
   d. 1.59 $m^2$

   Answer: _____

5. What is the recommended dose for a 15 kg child who is 165 cm tall if the dosing guidelines are 40 mg/$m^2$?
   a. 27.5 mg
   b. 33.2 mg
   c. 41.3 mg
   d. 42.4 mg

   Answer: _____

# Lesson 41 Content Check

1. Young's Rule is based on _____ whereas Clark's Rule is based on _____.

2. Use the West Nomogram to determine the surface area of the following children (who are all of normal height for their weight):

    a. 56 lbs

    Answer: _____

    b. 30 lbs

    Answer: _____

    c. 15 lbs

    Answer: _____

    d. 6 lbs

    Answer: _____

3. Calculate the BSA for a child with the following:

    a. Weight: 34 kg Height: 40 in

    Answer: _____

    b. Weight: 80 lb Height: 45 in

    Answer: _____

    c. Weight: 20 kg Height: 63 cm

    Answer: _____

    d. Weight: 15 kg Height: 30 in

    Answer: _____

e. Weight: 22 lb  Height: 24 in

Answer: _____

4. A patient is ordered acyclovir. They weigh 58.2 lbs and 200 cm tall. The reference text states that in children <40 kg, the dosing is 600 mg/m²/dose QID, and in children ≥40 kg, the dose is 800 mg QID. What is the appropriate dose for this patient?

Answer: _____

5. Answer the following questions for a 35 lb child who is 140 cm tall:

   Ordered: caspofungin

   Reference text: 70 mg/m²/dose on day 1, then 50 mg/m²/dose once daily

   a. How much medication should the patient get, in mg, on day 1?

Answer: _____

   b. How much medication should the patient get, in mg, on day 2?

Answer: _____

   c. How much medication, in grams, would the patient receive over a 7-day hospital stay?

Answer: _____

6. Answer the following questions for a 100 lb child who is 4' 6" tall:

   Ordered: allopurinol

   Reference text: 50 – 100 mg/m²/dose q8h; maximum daily dose of 300 mg/m²

a. What is the minimum dose required for this patient?

Answer: _____

b. What is the maximum dose allowed for this patient?

Answer: _____

7. Answer the following questions for a 11.5 lb infant who is 22.5 inches long:

Ordered: amiodarone

Reference text: loading dose of 347 – 462 mg/m²/day split into 2 doses; maintenance dose of 116 – 231 mg/m²/day administered once daily

a. What is the minimum loading dose required for this patient?

Answer: _____

i. What is the total daily loading dose that the patient will be getting if they are only given the minimum allowed?

Answer: _____

b. What is the maximum loading dose allowed for this patient?

Answer: _____

i. What is the total daily loading dose that the patient will be getting if they are given the maximum amount allowed?

Answer: _____

c. What is the minimum maintenance dose required for this patient?

Answer: _____

d. What is the maximum maintenance dose allowed for this patient?

Answer: _____

# LESSON 42

# Pediatric IV Administration

Volume control sets are specialized IV administration sets that contain a plastic, calibrated fluid chamber (sometimes called a burette) that is built into the line. These chambers can usually hold up to 150 mL of additional volume. They are often used as a replacement, or in favor of, a mini-bag or IVPB as they have an injection port for direct administration of medication into the existing IV line. They ensure exact, precise delivery of IV doses while also preventing air from entering the IV line, and they allow the primary IV fluid to continue once the chamber is empty, if necessary. Due to their use primarily in pediatric patients, volume control sets have a DF of 60.

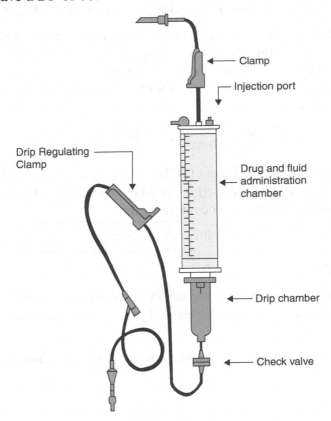

Figure 42.1 **Volume Control Set**

# Volume Control Set Calculations

Since medication is added directly into the chamber of a volume control set, the calculations will change slightly due to the context and use of these tools. Nurses will dilute the medication added to the chamber directly with the IV fluids that already exist in the primary bag hanging above the burette and tubing line.

> **Example:** Ordered: $D_{7.5}W$ IV at 40 mL/hr with azithromycin 100 mg IV qd to be infused over a 50 minute time frame with a volume control set.
>
> On hand: azithromycin 250 mg/5 mL

1. **How many mL of azithromycin should be added to the control chamber?**

    Determine the amount of drug the patient needs by using the ratio-proportion method.

    $$\frac{250 \text{ mg}}{5 \text{ mL}} = \frac{100 \text{ mg}}{x \text{ mL}} \quad \text{where } x = \boxed{2 \text{ mL}}$$

2. **How many mL of $D_{7.5}W$ should be added to the control chamber?**

    Notice that a flow rate is given in this problem, but the patient is only going to be using the medication for 50 minutes instead of the full hour, as indicated by the flow rate. So, the total volume of the fluid will not be 40 mL, but can be calculated by setting up a dimensional analysis problem starting with the given flow rate and incorporating the amount of time it will be used:

    $$\frac{40 \text{ mL}}{1 \text{ hr}} \times \frac{1 \text{ hr}}{60 \text{ min}} \times \frac{50 \text{ min}}{1} = 33.33 \text{ mL}$$

    To determine the amount of diluent (IV base) to add to the control chamber to mix with the medication that was added in the first step, subtract the volume of drug needed from the total volume to be administered.

    $$33.33 \text{ mL} - 2 \text{ mL} = \boxed{31.33 \text{ mL of } D_{7.5}W}$$

**STOP AND PRACTICE:**

1. A prescriber orders clindamycin 300 mg IV to be infused over a 30-minute time frame in $D_5NS$ at 50 mL/hr with a volume control set. The pharmacy has clindamycin 500 mg/5 mL on hand.

    a. How many mL of clindamycin should be added to the control chamber?

    Answer: _____

    b. How many mL of $D_5NS$ should be added to the control chamber?

    Answer: _____

## LESSON 42: Pediatric IV Administration

2. A prescriber orders ½ NS IV at 100 mL/hr with Bactrim® 400 mg BID to be infused over a 75-minute time frame with a volume control set. The pharmacy has Bactrim® 800 mg/5 mL on hand.

   a. How many mL of Bactrim® should be added to the control chamber?

   Answer: _____

   b. How many mL of ½ NS should be added to the control chamber?

   Answer: _____

3. A nurse receives an order for acetaminophen 100 mg IV Q6h in $D_{2.5}W$ at 20 mL/hr to be infused over a 15-minute time frame with a volume control set. The pharmacy has acetaminophen 160 mg/5 mL in stock.

   a. How many mL of acetaminophen should be added to the control chamber?

   Answer: _____

   b. How many mL of $D_{2.5}W$ should be added to the control chamber?

   Answer: _____

4. Ordered: $D_{10}W$ IV at 80 mL/hr with imipramine 25 mg IV qd to be infused over a 30-minute time frame with a volume control set. On hand: imipramine 50 mg/5 mL

   a. How many mL of imipramine should be added to the control chamber?

   Answer: _____

   b. How many mL of $D_{10}W$ should be added to the control chamber?

   Answer: _____

## Flushes

When using a volume control set, it is common practice for nurses to clear the IV line after use to ensure that all of the medication has been administered to the pediatric patient. They usually use the same diluent that the medication was mixed with, which is same as the solution in the primary IV bag. This process is called "flushing" the IV line. It can be done with a volume between 10 and 30 mL, and usually the flush is calculated into the flow rate in order to make administration as simple as possible. Always consult the policy and procedure manual of the institution when in doubt on the quantity of flushes used, or the calculations to consider.

# UNIT 7: Pediatric Pharmacy Math

> **Example:** Famotidine 20 mg IV q12h in 100 mL $D_5NS$ over a time period of 50 minutes using a microdrip set is ordered for a pediatric patient. The nurse is instructed to flush with 10 mL of $D_5NS$. The pharmacy stocks famotidine in a strength of 100 mg/5 mL.

1. **How many mL of famotidine should be added to the control chamber?**

   Determine the amount of drug the patient needs by using the ratio-proportion method.

   $$\frac{100 \text{ mg}}{5 \text{ mL}} = \frac{20 \text{ mg}}{x \text{ mL}} \quad \text{where } x = \boxed{1 \text{ mL}}$$

2. **How many mL of $D_5NS$ should be added to the control chamber?**

   Notice that a flow rate is not given in this problem, but the total volume of the fluids to be given is mentioned. To determine the amount of diluent (IV base) to add to the control chamber to mix with the medication that was added in the first step, subtract the volume of drug needed from the total volume to be administered.

   $$100 \text{ mL} - 1 \text{ mL} = \boxed{99 \text{ mL of } D_5NS}$$

3. **What is the flow rate in gtt/min?**

   Identify, or determine the flow rate in mL/min from the problem itself. Make sure you add the volume of the flush to the volume of the IV bag before you determine a flow rate! Also note that a microdrip set is 60 gtt/mL.

   $$\frac{100 \text{ mL(IV fluids)} + 10 \text{ mL(flush)}}{50 \text{ min}} \times \frac{60 \text{ gtt}}{1 \text{ mL}} = \boxed{\frac{132 \text{ gtt}}{\text{min}}}$$

## STOP AND PRACTICE:

1. A prescriber orders clarithromycin 200 mg IV q12h in 25 mL $D_5NS$ over 40 minutes using a volume control set and then states that the nurse should flush with 25 mL. The pharmacy has clarithromycin 250 mg/5 mL in stock.

   a. How many mL of clarithromycin should be added to the control chamber?

   Answer: _____

   b. How many mL of $D_5NS$ should be added to the control chamber?

   Answer: _____

   c. What is the flow rate in gtt/min?

   Answer: _____

## LESSON 42: Pediatric IV Administration

2. A prescriber enters the following order: $D_{20}W$ IV at 50 mL/hr with phenytoin 50 mg IV qd to be infused over a 30-minute time frame with a volume control set and then flushed with 15 mL. The nurses station stocks phenytoin in a 25 mg/mL solution.

   a. How many mL of phenytoin should be added to the control chamber?

   Answer: _____

   b. How many mL of $D_{20}W$ should be added to the control chamber?

   Answer: _____

   c. What is the flow rate in gtt/min?

   Answer: _____

3. An order comes across the nurse's station for 50 mg of diphenhydramine IV q8h in 30 mL $D_5NS$ over 30 minutes using a volume control set followed by a flush of 15 mL. The automated dispensing system in the nurse's station stocks ampules of diphenhydramine with a concentration of 100 mg/3 mL.

   a. How many mL of diphenhydramine should be added to the control chamber?

   Answer: _____

   b. How many mL of $D_5NS$ should be added to the control chamber?

   Answer: _____

   c. What is the flow rate in gtt/min?

   Answer: _____

# UNIT 7: Pediatric Pharmacy Math

## Sampling the Certification Exam:

1. A medication is prescribed to a child with orders to add 20 mg in $D_{2.5}W$ for a total of 45 mL over the course of 30 minutes using a volume control set. The nurse has also been instructed to follow this by a flush of 30 mL. The medication is available with a concentration of 50 mg/5 mL. How much drug should be added to the control chamber?

   a. 1 mL

   b. 2 mL

   c. 3 mL

   d. 4 mL

   Answer: _____

2. A medication is prescribed to a child with orders add 20 mg in $D_{2.5}W$ for a total of 45 mL over the course of 30 minutes using a volume control set. The nurse has also been instructed to follow this by a flush of 30 mL. The medication is available with a concentration of 50 mg/5 mL. How much $D_{2.5}W$ should be added to the control chamber?

   a. 43 mL

   b. 45 mL

   c. 47 mL

   d. 50 mL

   Answer: _____

3. A medication is prescribed to a child with orders to add 20 mg in $D_{2.5}W$ for a total of 45 mL over the course of 30 minutes using a volume control set. The nurse has also been instructed to follow this by a flush of 30 mL. The medication is available with a concentration of 50 mg/5 mL. What is the flow rate in mL/hr?

   a. 75 mL/hr

   b. 100 mL/hr

   c. 125 mL/hr

   d. 150 mL/hr

   Answer: _____

4. A medication is prescribed to a child with orders to add 20 mg in $D_{2.5}W$ for a total of 45 mL over the course of 30 minutes using a volume control set. The nurse has also been instructed to follow this by a flush of 30 mL. The medication is available with a concentration of 50 mg/5 mL. What is the flow rate in gtt/min if the volume control set is calibrated to 60.

a. 75 gtt/min

b. 100 gtt/min

c. 125 gtt/min

d. 150 gtt/min

Answer: _____

5. Volume control sets usually contain burettes that hold up to ____ mL of solution.

a. 100

b. 120

c. 150

d. 200

Answer: _____

## Lesson 42 Content Check

1. A prescriber orders baclofen 15 mg IV to be infused over a 20-minute time frame in NS at a rate of 100 mL/hr with a volume control set. The pharmacy has baclofen 10 mg/mL on hand.

   a. How many mL of baclofen should be added to the control chamber?

   Answer: _____

   b. How many mL of NS should be added to the control chamber?

   Answer: _____

   c. If this patient repeats this dose three times more throughout the day, how many milligrams of baclofen have they received?

   Answer: _____

2. A prescriber orders celecoxib 45 mg IV to be infused over a quarter of an hour in $D_{10}W$ at a rate of 50 mL/hr with a volume control set. The pharmacy has celecoxib 100 mg/5 mL on hand.

   a. How many mL of celecoxib should be added to the control chamber?

   Answer: _____

b. How many mL of $D_{10}W$ should be added to the control chamber?

Answer: _____

c. How many grams of celecoxib would the patient get if they had this medication dosed BID for a 3-day hospital stay?

Answer: _____

3. A prescriber orders 3.125 mg of carvedilol IV with an infusion time of half an hour in $D_5W$ at a rate of 60 mL/hr with a volume control set. The pharmacy has carvedilol 6.25 mg/mL on hand.

   a. How many mL of carvedilol should be added to the control chamber?

Answer: _____

   b. How many mL of $D_5W$ should be added to the control chamber?

Answer: _____

4. A prescriber ordered the following for a pediatric patient: levofloxacin 500 mg in $D_5$ 0.33% NaCl IV at 65 mL/hr to be hung for a period of 40 minutes using a volume control set and followed by a 15 mL flush. The automated dispensing machine stocks levofloxacin 50 mg/mL.

   a. How much levofloxacin will be added to the control chamber?

Answer: _____

   b. How much $D_5$ 0.33% NaCl should be added to the control chamber?

Answer: _____

   c. What is the flow rate in gtt/min if the volume control set is calibrated the same as a microdrip set?

Answer: _____

   d. If the nurse was instructed to repeat this dose for the patient once daily for a 5-day hospital stay, how many mL of levofloxacin would they get in total?

Answer: _____

## LESSON 42: Pediatric IV Administration

5. A prescriber orders the following for a pediatric patient: ceftriaxone 0.42 g IV in $D_5NS$ for a total volume of 30 mL; run for 30 minutes using a volume control set and flush with 15 mL. The pharmacy sends up ceftriaxone 500 mg/5 mL.

   a. How much ceftriaxone will be added to the control chamber?

   Answer: _____

   b. How much $D_5NS$ should be added to the control chamber?

   Answer: _____

   c. What is the flow rate in mL/hr?

   Answer: _____

   d. What is the flow rate in gtt/min if the volume control set is calibrated the same as a microdrip set?

   Answer: _____

6. The following orders are received by the pharmacy: clindamycin 285 mg IV in $D_5NS$ for a total volume of 45 mL; run for 60 minutes using a volume control set and flush with 15 mL. The nursing station stocks clindamycin 75 mg/0.5 mL.

   a. How much clindamycin will be added to the control chamber?

   Answer: _____

   b. How much $D_5NS$ should be added to the control chamber?

   Answer: _____

   c. What is the flow rate in mL/hr?

   Answer: _____

   d. What is the flow rate in gtt/min if the volume control set is calibrated the same as a minidrip set?

   Answer: _____

# LESSON 43

# Daily Maintenance Needs

Certain disease states and conditions, such as severe infection or burn, or when a patient is not allowed to have anything by mouth (i.e., "NPO" status), require the patient to be on maintenance fluids to prevent starvation and maintain their body composition while they are being treated for their ailment. These fluids are mostly water with electrolytes and trace elements added to keep the body in a safely hydrated and healthy condition and are calculated on a 24-hr basis, hence the name – daily maintenance fluids.

Nurses keep an eye on and document the volume of patient's intake (other fluids) and outtake (through urine and stool) but they cannot measure fluid loss from breathing, sweating, and other events that might occur (ex: emesis). Daily maintenance needs can help account for the loss of fluid that are not measurable.

The standardization of daily fluid needs, especially for children, takes into account their body weight (in kg) in the following way:

10 kg or less = _____ mL/kg

10 – 20 kg = 1,000 mL + _____ mL/kg over 10 kg.

Above 20 kg = 1,500 mL + _____ mL/kg over 20 kg.

Using the rules:

> **Example:** What is the daily maintenance fluids needed for a 13 kg child?

1. A 13 kg child meets the requirements of the middle rule as seen above. Therefore, the base need is 1,000 mL. To determine how many additional mL are needed, the weight needs to be adjusted to account for the amount above 10 kg:

$$13 \text{ kg} - 10 \text{ kg} = 3 \text{ kg}$$

2. Use the ratio-proportion method to determine how many additional mL are needed:

$$\frac{50 \text{ mL}}{1 \text{ kg}} = \frac{x \text{ mL}}{3 \text{ kg}} \quad \text{where } x = 150 \text{ mL}$$

3. Add the two steps together to determine the total amount needed:

$$1{,}000 \text{ mL} + 150 \text{ mL} = 1{,}150 \text{ mL of fluids needed}$$

Occasionally, a follow-up question of the flow rate will be asked. Since these medications are calculated on a 24-hr basis, the ratio-proportion method would be a simple set-up to find the solution, but be sure to watch the units that the question is asking for.

**Example:** What is the flow rate in mL/hr for the daily maintenance fluids needed for a 13 kg child?

$$\frac{1{,}150 \text{ mL}}{24 \text{ hr}} = \frac{x \text{ mL}}{1 \text{ hr}} \quad \text{where } x = 47.92 \text{ mL, so the answer would be } \boxed{47.92 \text{ mL/hr}}$$

**STOP AND PRACTICE:** Calculate the daily maintenance fluids for the following:

1. A child weighing 9 kg

   Answer: _____

   a. What is the flow rate in mL/hr for this patient?

   Answer: _____

2. A child weighing 8.3 lb

   Answer: _____

   a. What is the flow rate in mL/hr for this patient?

   Answer: _____

3. A child weighing 15 kg

   Answer: _____

   a. What is the flow rate in mL/hr for this patient?

   Answer: _____

4. A child weighing 27 lb

   Answer: _____

   a. What is the flow rate in mL/min for this patient?

   Answer: _____

5. A child weighing 54 lb

   Answer: _____

   a. What is the flow rate in mL/min for this patient?

   Answer: _____

# LESSON 43: Daily Maintenance Needs

## Sampling the Certification Exam:

1. What are the daily maintenance fluid needs for a child who weighs 5 kg?

    a. 5 mL
    b. 50 mL
    c. 500 mL
    d. 1,000 mL

    Answer: _____

2. What would the flow rate be for the daily maintenance fluid needs for a child weighing 18 lb?

    a. 21 mL/hr
    b. 30 mL/hr
    c. 34 mL/hr
    d. 41 mL/hr

    Answer: _____

3. What are the daily maintenance fluid needs for a child who weighs 21 kg?

    a. 1,500 mL
    b. 1,520 mL
    c. 1,750 mL
    d. 2,100 mL

    Answer: _____

4. What would the flow rate be for the daily maintenance fluid needs for a child weighing 49 lb?

    a. 51 mL/hr
    b. 52 mL/hr
    c. 62 mL/hr
    d. 64 mL/hr

    Answer: _____

5. What are the daily maintenance fluid needs for a neonate who weighs 4 lb, 3 oz?
   a. 100 mL
   b. 120 mL
   c. 150 mL
   d. 190 mL

   Answer: _____

## Lesson 43 Content Check

1. What are the daily maintenance fluid needs for a 16 lb child?

   Answer: _____

   a. What is the flow rate of this solution in milliliters per hour?

   Answer: _____

   b. What is the flow rate of this solution in gtt/min if an administration set with a DF of 20 is used?

   Answer: _____

2. What are the daily maintenance fluid needs for a 5200 g infant?

   Answer: _____

   a. What is the flow rate of this solution in milliliters per minute?

   Answer: _____

   b. What is the flow rate of this solution in gtt/min if an administration set with a DF of 10 is used?

   Answer: _____

3. What are the daily maintenance fluid needs for a 7.9 kg child?

   Answer: _____

   a. What is the flow rate of this solution in drops per hour using a microdrip tubing set?

   Answer: _____

   b. What is the flow rate of this solution in gtt/min if an administration set with a DF of 20 is used instead?

   Answer: _____

4. What are the daily maintenance fluid needs for a 44 lb child?

    Answer: _____

    a. What is the flow rate of this solution in drops per minute using a pediatric administration set?

    Answer: _____

    b. How many mL per pound is being administered for this child?

    Answer: _____

5. What are the daily maintenance fluid needs for an 85 lb child?

    Answer: _____

    a. What is the flow rate of this solution in milliliters per hour?

    Answer: _____

    b. What is the flow rate of this solution in gtt/min if an administration set with a DF of 15 is used?

    Answer: _____

## Unit 7 Content Review

**Complete each statement.**

1. Body surface area (BSA) is expressed as _____.
2. The two methods previously used to calculate a pediatric dose that are no longer utilized are _____'s Rule and _____'s Rule.
3. Pharmacy technicians can calculate the BSA by using a chart called a _____.

**True/False** - *Indicate whether the statement is true or false.*

4. Young's Rule is based on age of the child, not weight, compared to an adult dose. **T / F**
5. The use of Clark's rule to calculate pediatric dosages is much less accurate than other pediatric methods. **T / F**
6. Volume control sets for pediatric patients are microdrip sets that use a drop factor of 60 gtt / mL. **T / F**

## UNIT 7: Pediatric Pharmacy Math

**Numeric Response** - *Convert the pediatric weights from pounds to kilograms.*

7. 25 lb

   Answer: _____

8. 82 lb

   Answer: _____

9. 7 lb 4 oz

   Answer: _____

10. 19 lb 15 oz

    Answer: _____

11. 32 lb 6 oz

    Answer: _____

**Calculate the amount of the drugs based on the child's body weight (in kilograms).**

12. Ordered: oseltamivir 4 mg / kg po
    On hand: oseltamivir 60 mg per 5 mL
    How many milliliters would a child weighing 92.4 lb receive?

    Answer: _____

13. Ordered: cefpodoxime 2.5 mg / kg
    On hand: cefpodoxime 50 mg per 5 mL
    How many milliliters would a child weighing 55 lb receive?

    Answer: _____

14. Ordered: Humulin R® insulin 0.14 U / kg SC bid. ac breakfast and dinner
    If the weight of a child is 35 kg, how many units would they receive in one dose?

    Answer: _____

**LESSON 43:** Daily Maintenance Needs 447

15. Ordered: fentanyl 3.3 mcg per kg IM

    On hand: fentanyl 50 mcg per mL

    How many milliliters would you give to a child whose weight is 66 lb?

    Answer: _____

16. Ordered: 1% lidocaine 0.5 mg / kg IV bolus

    How many milligrams would you give to a child whose weight is 45 kg?

    Answer: _____

**Calculate the daily maintenance fluids for pediatric patients.**

17. Calculate the total volume and hourly IV flow rate for a 28.6 lb child receiving maintenance IV fluids.

    Infuse _____ mL at _____ mL / hr

18. Calculate the total volume and hourly IV flow rate for a 55 lb child receiving maintenance IV fluids.

    Infuse _____ mL at _____ mL / hr

19. Calculate the total volume and hourly IV flow rate for a 7.7 lb neonate receiving maintenance fluids.

    Infuse _____ mL at _____ mL / hr

**Calculate the BSA (in m$^2$) value of the following:**

20. Height: 60 cm  Weight: 6 kg

    Answer: _____

21. Height: 68 in  Weight: 170 lb

    Answer: _____

22. Height: 100 cm  Weight: 17 kg

    Answer: _____

23. Height: 64 in  Weight: 63 kg

    Answer: _____

24. Height: 164 cm  Weight: 58 kg

    Answer: _____

25. Ordered: Vancocin® 7.5 mg IM q12 h

    On hand: Vancocin® multidose vial, 20 mg / 2 mL. According to the package insert, a full-term neonate, up to 1 week of age, may be administered up to 4 mg / kg / day in 2 equal doses every 12 hours. This patient is a 5-day-old newborn who weighs 7 lb 12 oz.

    a. Is this dosage safe?

    Answer: _____

    b. How much drug should be administered for each dose?

    Answer: _____

26. Ordered: cefaclor susp 270 mg po qd 1 h ac

    On hand: cefaclor susp 90 mg per 5 mL.

    According to the package insert, the usual recommended daily dosage for children who weigh 45 kg or less is 9 mg / kg qd. The patient weighs 66 lb.

    a. Is this dosage safe?

    Answer: _____

    b. How much drug should be administered for each dose?

    Answer: _____

27. Ordered: digoxin 125 mcg IVPB over 30 minutes

    On hand: digoxin 300 mcg / 1 mL single-use ampule

    According to the package insert, the recommended daily starting dose is 5 mcg / kg / day, administered as a single daily injection by SC bolus injection, by short IV infusion (15 to 30 minutes). Note: The child weighs 25 kg.

    a. Is this dosage safe?

    Answer: _____

    b. How much drug should be administered for each dose?

    Answer: _____

28. Ordered: divalproex syrup 100 mg po q12 h

    On hand: divalproex syrup 250 mg / 5 mL

    According to the package insert, the initial daily dose for pediatric patients is 15 mg / kg / day. Patient is 14 lb.

    a. Is this dosage safe?

    Answer: _____

    b. How much drug should be administered for each dose?

    Answer: _____

# Appendix A – Math Concepts

| Step 1: Identify the type of problem ||| Step 2: Determine the method you will use to solve the problem. | Step 3: Set up the problem. What to remember: |
|---|---|---|---|---|
| Unit/ Concept | Type | Key Words | | |
| Basic Math Calculations | Fractions | Fraction | Follow Rules | KCF to divide fractions |
| | Decimals | Decimal | Follow Rules | Place value, Leading zeros |
| | Rounding and Place Value | Round, Place Value | Follow Rules | 4 or less, keep; 5 or more, round up |
| | Ratios | Ratio | Follow Rules | Most are 1: x |
| | Proportions | — | Cross multiply and divide | Must know 3 out of the 4 positions |
| | Percent's | Percent | Ratio/Proportion or Dimensional Analysis | x % = x / 100 |
| Conversions | Roman / Arabic Numerals | Roman, Arabic | Follow rules | Value of each numeral |
| | Temperature | Fahrenheit, Celsius | Temperature Formula | 9C = 5F - 160 |
| | Military Time to Standard AM/PM | Military, AM/PM | Follow rules | 0000 – 2359; PM = time + 1200 |
| | Metric | Kilo-, gram/ Liter, milli-, micro- | KSMM Method | Cannot convert weight to volume; standard weight = g, standard volume = L; prefixes are 1,000 degrees of separation |
| | Household / Apothecary | Gallon, quart, pint, cup, ounce, tablespoon, teaspoon | Ratio/Proportion or Dimensional Analysis | 328224, order of units, meaning of abbreviations |
| | Other | — | Ratio/Proportion or Dimensional Analysis | 1 tsp = 5 mL, 1 kg = 2.2 lb., U-# insulin means that 1 mL equals the # units |

*(continued)*

# Appendix A – Math Concepts

| Step 1: Identify the type of problem ||| Step 2: Determine the method you will use to solve the problem. | Step 3: Set up the problem. What to remember: |
|---|---|---|---|---|
| Unit/ Concept | Type | Key Words | | |
| Prescription Processing Calculations | Dosage | Dose, Dosage | Ratio/Proportion or Dimensional Analysis | Always compare what you need (unknown) to what you have (known) |
| | Body Surface Area | BSA | BSA Formula | BSA (m²) = $\sqrt{\dfrac{Wt\,(kg) \times ht\,(cm)}{3600}}$ |
| | Quantity | How much/how many | Ratio/Proportion or Dimensional Analysis | Quantity sufficient = Dose at one time x doses per day x days indicated |
| | Day Supply | How long/how many days | Ratio/Proportion or Dimensional Analysis | Day supply = Quantity /(dose at one time x doses per day) |
| | Insurance | Insurance | Ratio/Proportion or Dimensional Analysis | Terminology (copay, "cash" pay, premium, deductible, AWP, U&C, MAC, PBM) |
| | Compounding | To make/to compound | Ratio/Proportion or Dimensional Analysis | Always compare (one ingredient at a time) what is needed to the Master Formula Record and quantity that it makes |
| Specialty Calculations | Percent Error | Percent error | Formula | Error = quantity desired – quantity actual OR sensitivity of machine being used. Percent error = (E x 100%)/quantity desired |
| | Concentration / Dilution Calculations | How many mL/ How much = alligation method. What percent/ what is the concentration = dilution [ratio/proportion] | Alligation Method or Ratio/Proportion | Always use % - convert to a % if not given that way; w/v is always g/mL, v/v is always mL/mL, w/w is always g/g; percent strength = x/100 |

*(continued)*

# Appendix A – Math Concepts

| Step 1: Identify the type of problem ||| Step 2: Determine the method you will use to solve the problem. | Step 3: Set up the problem. What to remember: |
| Unit/ Concept | Type | Key Words | | |
| --- | --- | --- | --- | --- |
| Specialty Calculations | Powder Volume / Reconstitution Calculations | Powder volume | Powder volume formula and Ratio/ Proportion | PV = FV – D; Powder volume and weight cannot change, but diluent and final volume can; concentration and amount of diluent have an inverse relationship |
| | IV Flow Rate Calculations | Infusion rate/drip rate/flow rate | Dimensional Analysis | If given weight, that is a good place to start; watch the units given vs. the units asked for! |
| Business Math Calculations | Income, Overhead, Profit (Gross vs. Net) | Income, overhead, profit, gross profit, net profit, expenses, percent profit, sales | Ratio/Proportion | Profit ($ or %) = Income ($ or %) – Overhead ($ or %)* (Stay consistent with units) Gross Profit (GP) = sales price – acquisition cost Net Profit (NP) = GP – dispensing fee Dispensing Fee = total pharmacy overhead / total pharmacy prescriptions Always compare to overhead = 100% |
| | Mark-up vs Discount | Mark-up, discount, sales price, acquisition cost | Ratio/Proportion or Dimensional Analysis | Always compare to sales price = 100% |
| | Depreciation | Depreciation | Depreciation Formula | Depreciation = Total cost – disposal value / estimated life in years |
| | Inventory Calculations | Stock, turnover, BOH, PAR | Turnover Formula, Follow Rules, Ratio/ Proportion | Turnover rate = Total purchases for a given time / Value of inventory at that time |

# Appendix B – Visual dosage chart for 30 days

|        | 1st dose | 2nd dose | 3rd dose | 4th dose | 5th dose | 6th dose |
|--------|----------|----------|----------|----------|----------|----------|
| Day 1  |          |          |          |          |          |          |
| Day 2  |          |          |          |          |          |          |
| Day 3  |          |          |          |          |          |          |
| Day 4  |          |          |          |          |          |          |
| Day 5  |          |          |          |          |          |          |
| Day 6  |          |          |          |          |          |          |
| Day 7  |          |          |          |          |          |          |
| Day 8  |          |          |          |          |          |          |
| Day 9  |          |          |          |          |          |          |
| Day 10 |          |          |          |          |          |          |
| Day 11 |          |          |          |          |          |          |
| Day 12 |          |          |          |          |          |          |
| Day 13 |          |          |          |          |          |          |
| Day 14 |          |          |          |          |          |          |
| Day 15 |          |          |          |          |          |          |
| Day 16 |          |          |          |          |          |          |
| Day 17 |          |          |          |          |          |          |
| Day 18 |          |          |          |          |          |          |
| Day 19 |          |          |          |          |          |          |
| Day 20 |          |          |          |          |          |          |
| Day 21 |          |          |          |          |          |          |
| Day 22 |          |          |          |          |          |          |
| Day 23 |          |          |          |          |          |          |
| Day 24 |          |          |          |          |          |          |
| Day 25 |          |          |          |          |          |          |
| Day 26 |          |          |          |          |          |          |
| Day 27 |          |          |          |          |          |          |
| Day 28 |          |          |          |          |          |          |
| Day 29 |          |          |          |          |          |          |
| Day 30 |          |          |          |          |          |          |

# Appendix C – Visual dosage chart for 4 weeks

|        | Monday | Tuesday | Wednesday | Thursday | Friday | Saturday | Sunday |
|--------|--------|---------|-----------|----------|--------|----------|--------|
| Week 1 |        |         |           |          |        |          |        |
| Week 2 |        |         |           |          |        |          |        |
| Week 3 |        |         |           |          |        |          |        |
| Week 4 |        |         |           |          |        |          |        |